INTRODUCTION TO PATENTS INFORMATION

2nd edition

Prepared by PATENTS INFORMATION STAFF

Edited by STEPHEN VAN DULKEN

THE BRITISH LIBRARY

Published by:

The British Library, Science Reference and Information Service, 25 Southampton Buildings, London WC2A 1AW.

First edition published in 1990.

For further information on SRIS publications contact Paul Wilson on 071-323 7472.

Design and desktop publishing by Tony Antoniou.

British Library Cataloguing in Publication Data

Dulken, Stephen Van
  Introduction to Patents Information. —
  2 Rev. ed. — (Information in Focus Series)
  I. Title II. Series
  025.06608741

  ISBN 0-7123-0790-7

Contents

PREFACE TO THE SECOND EDITION

The British Library Science Reference and Information Service (SRIS) is the United Kingdom's national centre for patents information. It holds one of the largest industrial property literature collections in the world and offers many patent information services including document supply, online searching, publications and training courses on the use of patents as a source of information.

SRIS originally published this guide to fill a clear need for a practical and readable primer to patents as a source of technical information. As far as we are aware, nothing comparable had ever been published. The good response that we have received has led us to bring out a second edition.

The guide is primarily for those who are either totally new to patents, or who have not yet fully mastered the basics of how to approach patents. At the same time, we hope that it will also be useful for those who are more advanced in their knowledge.

Introduction to patents information was originally meant as a coursebook for SRIS' patents seminars of the same name. These origins are reflected in a certain bias towards British interests, but patent documentation for the United States, Germany and Japan are all discussed, and this book should be useful for patents users abroad as well as in Britain.

The guide is divided into three main sections. The first explains the main points of the patent documentation of six patent offices: Britain, the European Patent Convention, the Patent Cooperation Treaty patents, the United States, Germany and Japan.

The second section covers the approaches to actually getting information from patents, such as going online, or using the published material from secondary services. There is also information on using material held at SRIS or at the 13 patent collections situated elsewhere in the UK.

The third section consists of appendices. These are a glossary, a bibliography, and lists of the information codes used by most countries in their documentation.

This new edition has been revised to include extra information and recent changes. In addition, the bibliography has been extensively revised, and citations to articles and brochures have been added at the end of the chapters. Use has also been made for the first time of tinted boxes containing useful supplementary information.

Little attempt has been made to cover anything other than the current documentation, since it is assumed that users will be mainly interested in recent documents.

Material for this edition was written by David Newton, Fiona Draper and Ian McKevitt as well as the editor. Desktop publishing was by Tony Antoniou.

Every effort has been made to make this publication accurate and up to date at the time of writing, but SRIS cannot accept any responsibility for any errors.

Suggestions for alterations or additions for future editions would be welcome. Send your suggestions to: Stephen van Dulken, SRIS, 25 Southampton Buildings, London WC2A 1AW. Tel 071 323 7916.

1 INTRODUCTION TO PATENTS INFORMATION

Advantages of patents Why should you use the million patent documents published annually as a source of information?

There are a number of advantages in using patents information:

- Currency of data. The publication of a patent application is often the first time that the information has ever been published. This is because the details of an invention have to be kept secret before an application is submitted to the Patent Office. Examples are Hollerith's patent of 1889 for punched cards, which was not shown to the public until 1914, and Watson Watt's 1936 radar patent. This was supposed to be secret during World War II, yet it was possible to look at, or to purchase a copy of the patent.

- Exclusivity of information. It has been estimated that 85% of the information in patents is never published anywhere else.

- Full and practical descriptions. A patent specification must have sufficient detail in text and illustrations so that an expert in the same branch of industry could recreate the invention.

- Ease of comparison. Patents have become increasingly standardized in their layout, although there are exceptions. This is useful since it saves the time of anyone who needs to look at numerous patents, or who is looking at foreign language patents.

- Availability of translations. An application must be made in every country in which protection is required. This means that there may be many English language specifications which are in effect translations of the technical content of foreign language patents.

- Supplementary information. Many patents are published with search reports prepared by the patent offices, listing patents and any other literature which were found in the literature search on the subject matter of the invention. This may be interesting in itself, and can be important in trying to determine if a published application is likely to succeed.

Uses of patents

By examining the advantages we can see a number of possible uses for patents information:

- Current awareness. Since patents are often the first or even the only source of information on a technological advance they are an essential element of any current awareness effort.

- Avoiding infringement. The patent literature on any topic where manufacturing or importing is contemplated should be studied to avoid infringing patents that are still protected in the country.

- Inspiration. Browsing through the patents on a subject of interest can encourage interesting ideas, particularly as it is often possible to find the same concepts being used in unrelated industries.

- Licensing opportunities. Even if a patent is still protected in the country it may be possible to negotiate a licence for its manufacture or import.

- Preliminary to Research and Development. A search through the patents literature should always be done at the start of Research and Development to avoid wasteful duplication. The idea may be protected by a patent, or that patent may have passed out of protection so that a detailed description is ready for use.

- Information on competitors. Checking the current patent literature increases awareness of what competitors are doing.

- Trends in technology. The patent classification can be used to plot technological trends with a simple online search for later analysis.

Patent specifications and the gazettes and indexes associated with them are used by a broad cross section of the community.

This includes inventors and companies, hoping to secure a temporary monopoly for a new idea; patent agents and searchers, carrying out infringement searches, to ensure that an apparently new idea is not blocked for use by a patent still in force; researchers generally, looking to see if an idea has ever been patented; the legal profession, gathering evidence for a court case involving patent rights; those looking for a solution to a technical problem; and anyone interested in what has been done in the past, or currently, in a particular topic.

What is a patent?

A patent is a contract between the state and the applicant by which a temporary monopoly is granted in return for disclosing all details of the invention. This is a simplification: what happens is not that the applicant is given a monopoly but rather that no one else can exploit that invention. This is because a patent may be useless except when combined with another, possibly patented, idea.

Criteria for the grant of patents

The word 'patent' is often used to describe any patent document. Correctly it ought only to be used for a document to which rights are, or have been, given. 'Specification' can be used to describe any document going through the patent system.

In order for a patent application to become a valid patent it must meet several criteria:

- It must be 'novel'. This means that it must be original as a patent, and indeed new in any published format. In addition, the application is unlikely to be accepted if the applicant describes it (except in confidence), or if it is manufactured before the application is submitted, so it is vital that it is kept secret until this time. Otherwise the invention is open to anyone to manufacture it.

- It must not be obvious. This means that it must not be a predictable improvement of something already in existence or described in the published literature. Theoretically if an uninventive person who knows all prior art thinks that an idea is an inventive step, then it is not obvious.

- It must be useful. It must do or be something of practical benefit, rather than being a scientific observation, or a work of art.

- It must be capable of being industrially reproduced. This criterion is applied very loosely: many chemical patents, for instance, refer to substances that cannot be reproduced in a factory environment.

- It must not be illegal or immoral. Examples of illegal patents would be ones for mantraps or counterfeiting machinery. However, some countries allow 'illegal' patents if the applicant intends to export the product to countries where they are legal.

- It must be detailed. The patent must be detailed enough so that someone skilled in the art can reconstruct the invention. This is fundamental, and failure to give sufficient detail can be cause to refuse a patent.

Some categories of invention are generally considered unpatentable. These include computer software and higher life forms (such as the genetically altered mouse), although the latter at least may soon become patentable.

Some patent offices publish specifications for utility models, sometimes called 'petty patents'. These are specifications where a smaller inventive step is required than for patents. They are generally for mechanical or electrical inventions. Germany and Japan are the principal countries to publish utility models.

Basic patenting procedure

The basic procedure of applying for patents is regulated by the Paris Convention for the Protection of Industrial Property of 1883, to which most countries belong.

An applicant for a patent draws up, usually with the help of a patent agent, a patent specification. This consists of a description, drawings, and the claims to the monopoly requested. It is then. together with the relevant forms, filed at (normally) the applicant's national patent office.

This initial, 'priority' application for a patent is in a sense provisional evidence of novelty. This is because the invention may turn out not to be new, or only an obvious improvement. Provided it is acceptable it is this filing date which is used to establish which of two rival patent applications is the first, or to establish that the applicant does have a claim to a monopoly right.

A filing or application number is also given to the application at this stage. This is the priority number which can be used to bring together all subsequent filings of the same invention in other patent offices, since most countries print the priority country, date and number of an invention on their patent documents.

This concept of novelty applies across the world and not just in the home country. Hence a British application on 1 July would be an anticipation of a French application on 2 July.

Should the applicant want protection outside the home country this must be applied for separately in each case. Under the Paris Convention these applications must be made within 12 months of the priority date. Most applicants wait the full 12 months until they actually file abroad.

Some companies, however, wait until almost the initial publication of the specification before filing abroad. This is called 'filing outside the Convention'.

In some countries, the applicant may be asked after filing the specification for a fee so that a search for anything that suggests that the invention is not new or is only obvious can be carried out. Other countries wait for the specification to be published before doing so, while other, usually smaller countries do not carry out a search at all.

Many industrialised countries publish applications 18 months after the priority date. The United States is a notable exception. The published foreign applications (and any subsequent published patents) are called the equivalents of the original application and together they form the patent family. Publication of the application does not imply that the patent office thinks that the invention is new. Its purpose is to allow others to note the existence of this invention.

Each patent office decides without consulting other offices whether or not to grant a patent. This is based on considering the claims, even if additional novel information is in the description. On the other hand, that additional novel information could be used to show that the claims of another patent application were not novel.

A number of countries do not accept patents for pharmaceutical products or processes. They should not, however, discriminate against foreigners in applying these policies.

If an application is considered acceptable then it is published a second time, usually in modified form, as a granted patent. Some countries have an opposition period during which the patent can be opposed on the grounds of non-novelty by interested parties. Germany is an example. Others, like Britain, grant the patent when the specification is published the second time but allow others to claim the patent is invalid and to instigate proceedings against it at any time. A successful attack on the patent would mean its revocation as if it was never protected at all.

In most countries it is necessary to pay, at intervals, renewal or maintenance fees to the Patent Office to keep the patent in force. Failure to pay the fees means that the invention is open to anyone else to manufacture or import the invention in that country.

Otherwise the patent runs its term and 'expires'. The full term varies from country to country but 20 years from the date of filing in that country is an increasingly common norm. A patent cannot be re-registered.

Layout of patents

The different parts of a patent specification are as follows:

Heading/ bibliographical data. In the same way that a book will have a title page, patents have a front page which gives useful bibliographical details. An abstract (compiled by the applicant) and, if appropriate, an illustration is usually given as well. Although this layout is becoming increasingly standardized the information given can vary from country to country. For example, British patents do not give the address of the inventor, only of the applicant (who may be the same person, of course).

Opening statement

Background information. This is often interesting and American patents are actually required to discuss the state of the art.

Problems. The nature of the technical problem is often outlined.

Description of the invention. The description explains the inventive step and how it works, with reference to illustrations.

Claims. This covers the legal aspects of the monopoly. Applicants like the claims to cover as much territory as possible but the examiner may force a modification of a claim if the area of monopoly seems unjustifiably large. The first, comprehensive claim describes the inventive step of the patent and it is followed by as many claims as are necessary to describe the different aspects of the inventive step.

Illustrations

Search report. This can vary in its detail but will at a minimum consist of a list of patent or other documents suggesting that the invention is not new. More detailed search reports would indicate which claims in the application were affected, the kind of relevance (mainly novelty or obviousness), and the exact page and line numbers thought to be relevant in the cited document.

The order in which the illustrations or search report are placed can vary from country to country, with some for instance putting the illustrations before the description of the invention.

International agreements have also led to standardized country codes, and INID (International agreed Numbers for the Identification of Data) codes for the bibliographic details on the front page. The codes are listed at page 115.

There has also been standardization in classification. Nearly all countries now use the International Patent Classification, or IPC, on their patents. This helps subject searching on an international basis using an online database.

Further reading

The scientific and technical information contained in patent specifications: the extent and time factors of its publication in other forms. F. Liebesny et al. *Information Scientist*, Dec 1974, 8 (4), 165-177.

The overlap of US and Canadian patent literature with journal literature. J. Allen, C. Oppenheim. *World Patent Information*, Oct 1979, 1 (2), 77-80.

Claim drafting: the British tradition. M.J. Daley. *CIPA*, 1981, Supplement, 1-9.

Patent protection for biotechnological inventions: a review of European and US practice. K. Percy. *World Patent Information*, 1989, 1 (3), 134-138.

The importance of the technological activities of the world's largest firms. P. Patel. *World Patent Information*, 1990, 12 (2), 89-94 [Mainly based on United States patents].

The patentability of computer-related inventions in the United Kingdom and the European Patent Office. B. Sherman. *European Intellectual Property Review*, March 1991, 13 (3), 85-94.

The value of patents. D. Needle, J, Needle. *Patent World*, Sept 1991, Issue 35, 20-36 [Contains charts showing the numbers of patents per million population for several countries as compared to output].

2 BRITISH PATENT PUBLICATIONS

Numeration

Between 1916 and 1 June 1978 British specifications were numbered by the Patent Office consecutively from 100,001 to about 1,600,000. They were 'sealed', or granted, after publication if fees were paid. A few publications from before that date are still being published under the old format and numeration.

Since 1 June 1978 applications are generally published 18 months from the priority date (there can be exceptions) in a sequence beginning with 2,000,001. On grant they are republished with the identical publication number. The first application to be published in 1991 was numbered 2,232,862.

Under the old and the new law patent applications were given a number like 9103456 when they were filed at the Patent Office. The first two digits indicate the year of filing and the remaining five a number from a sequence that begins from 1 each year. No specifications are numbered by these filing numbers.

Patenting procedure

The current British patenting procedure is illustrated on page 8. This procedure was laid down by the 1977 Patents Act.

British law requires British residents to file first at the British Patent Office even if no British patent is required. This is because the Ministry of Defence will intervene if the invention appears to be of a militarily sensitive nature. Some applications are as a result held in limbo, or are bought out by the government on a compulsory basis.

Otherwise, brief details are published in the *Official Journal (Patents)* or *OJ(P)*, about five weeks after filing. These details consist of applicant's name, title, filing date and number, and any priority details. The title may be vague, but a clearer title may be required when it is published. Britain is unusual in releasing information so quickly: in many countries it is necessary to wait until the application (or granted patent) is published.

A 'preliminary examination and search' of the patent is carried out. The preliminary examination is to see if the application has been set out correctly, and if the proper fees have been paid. If necessary the applicant will be told to reapply with a correct version within 12 months of the original application.

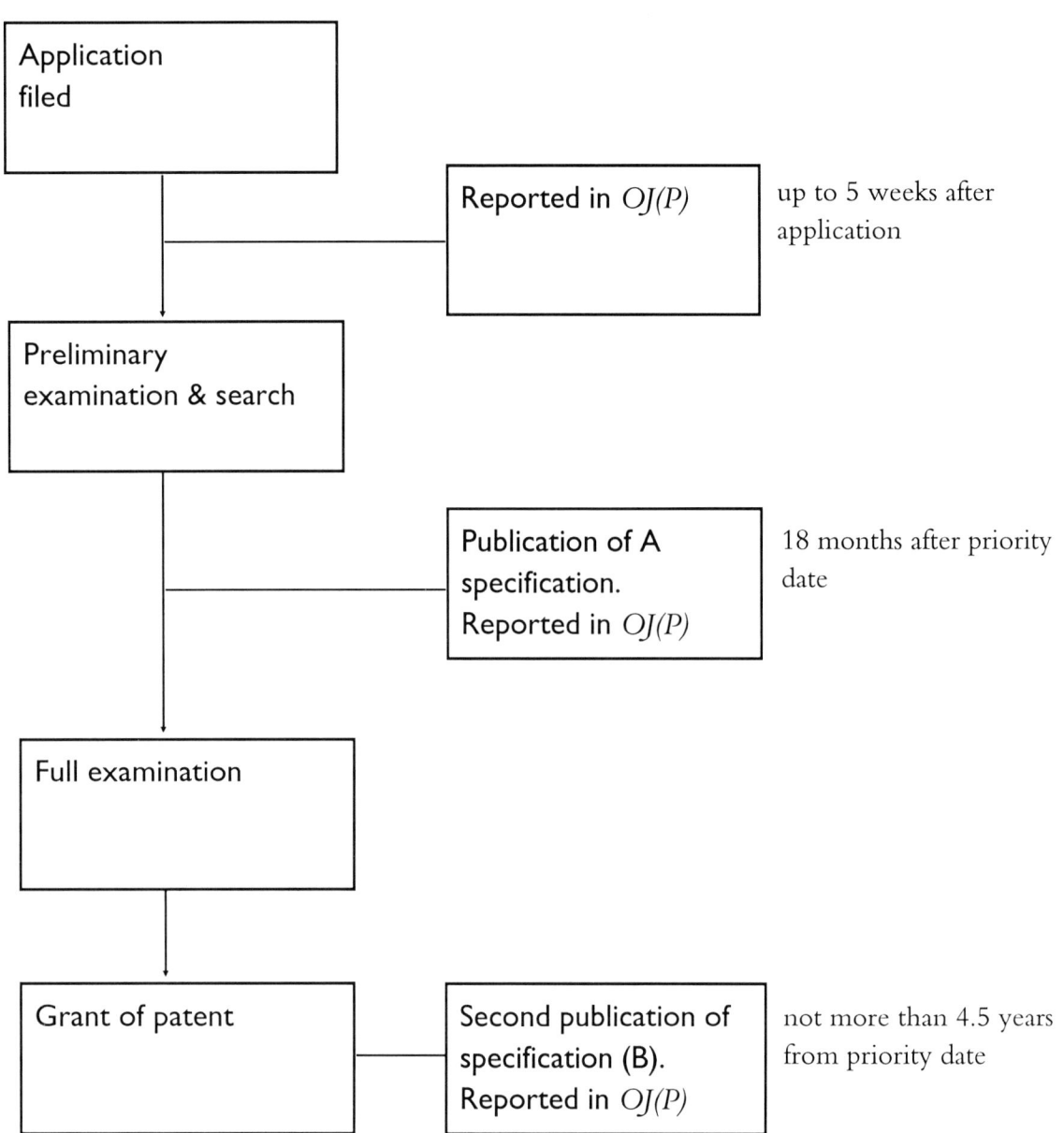

Application
filed

Reported in *OJ(P)*

up to 5 weeks after application

Preliminary
examination & search

Publication of A
specification.
Reported in *OJ(P)*

18 months after priority date

Full examination

Grant of patent

Second publication of
specification (B).
Reported in *OJ(P)*

not more than 4.5 years from priority date

OJ(P) = Official Journal (Patents)

Progress of a UK patent application under the Patents Act 1977

The applicant may also be told by the Patent Office to divide the application into two or more patent applications, or to merge two or more patent applications into one.

The search consists of the examiner looking for patents or other published material which suggest that the invention in question is not new, or only an obvious improvement. A detailed search report is sent to the applicant, who may withdraw the application before publication because of an unfavourable report. They can also withdraw the application for other reasons if they wish. The Patent Office will not actually reject any applications at this stage because of the search report. If the application is not published then it and its associated file remains confidential and is destroyed five years after the date of filing.

Patent applications are published weekly, normally on Wednesday. The week's applications are arranged in British classified order to assist current awareness searchers. The application includes a list of citations from the search report. From 1 April 1992 the complete search report, simiar to the European reports described on page 29, will be published with the application. The patent application is published about 18 months after the priority date, and hence perhaps only six months after the date of filing if it was originally applied for abroad. On the rare occasions when publication is delayed much beyond the 18 month period the invention is usually of a militarily sensitive nature. Again the *OJ(P)* reports on the publication.

At this stage the patent is often called an unexamined, or 'A' specification. It is not yet legally protected.

The applicant then has six months in which to apply for a substantive examination, or it is deemed to have been withdrawn. Substantive examination involves the Patent Office examiner taking into account the citations in the search report plus the criteria needed for a patent to be valid, such as being useful, and deciding whether or not to grant a patent. Interested parties may also make objections to the application. The search report may suggest that part or all of the patent is not new, or is merely obvious, and the examiner may refuse the application, or insist on its modification. The claims in particular may need to be modified, or reduced in number.

If the patent is accepted, then it is published again, if necessary in modified form, as a 'B', or granted specification. Again it is reported in the *OJ(P)*. The aim is to try to complete the procedure in four and a half years from the priority date.

Sometimes the granted patent is later attacked in court by another person or corporate body. This can take place at any time after grant. If the action is partially successful then the revised or deleted claims are given on an additional 'C' document. A totally successful attack means that the patent will be revoked.

Legal protection for the patent is 20 years from the date of filing in the UK, subject to renewal fees being paid annually (which is from five years of filing in Britain). If the renewal fees are not paid then it lapses, while if it runs the full term then it is said to have expired. The patent must be 'worked' within three years of grant or compulsory licensing to anyone interested will be forced on the applicant.

About 14,000 patent applications are published annually, of which a third are from foreigners. Foreign applicants are increasingly using the European Patent Convention to protect their inventions in Britain which is why the number of British applications has fallen considerably since the 1970s.

Specifications

Page 11 shows the front page of British patent application GB 2208189A. That is the correct citation: GB is the recognized country code for the United Kingdom, 2208189 is the number given to the published document, and A shows that it is only an application.

The numbers in brackets are called INID codes. Besides helping to structure the bibliographic details, they are useful if the patent is in a foreign language, since e.g. code 72 would always mean the name of the inventor.

At the top right we are given the date of publication, 8 March 1989. On the left appears code 21, with the application number, 8816039.5. The number after the full stop, 5, is a check digit for computer purposes. Code 22 gives the filing date.

The priority data takes up codes 31 to 33. From left to right they are the application number, the priority filing date, and the place where the priority filing was made, which had the country code JP, or Japan. The dates indicate that it was filed nearly a year later in Britain, and published about six months after that, in accordance with normal procedure.

Below the priority data is given the name and address of the applicant, Toshiba, followed by the inventor in code 72. The difference between the inventor and the applicant is that it is the latter who intends to 'work' the invention. In addition, only the address of the applicant is given, although if the invention is from a private inventor then the name will be repeated. Code 74 gives the name of the patent agent.

The classification details given by the Patent Office are on the right hand side. Code 51 gives the International Patent Classification (IPC) number, the superscript 4 showing that the fourth edition was used to classify the patent. Code 52 gives the British classification (Classification schemes are described in Chapter 8). Britain and the United States are the only major countries to use national classifications as well as the IPC.

Code 56 gives the results of the search report. In this case two patents were found to be relevant. Finally code 58 tells us what was searched: British classification heading H1Q plus the IPC subclass H01Q.

In the bottom half of the page there is a title and an abstract, both provided by the applicant, and a representative drawing.

It can be seen that the front page gives a great deal of useful information. (There are, incidentally, no copyright restrictions in photocopying patents.)

(12) UK Patent Application (19) GB (11) 2 208 189 (13) A

(43) Date of A publication 08.03.1989

(21) Application No 8816039.5

(22) Date of filing 06.07.1988

(30) Priority data
(31) 62169057   (32) 07.07.1987   (33) JP

(71) Applicant
Kabushiki Kaisha Toshiba

(Incorporated in Japan)

72 Horikawa-cho, Saiwai-ku, Kawasaki-shi,
Kanagawa-ken, Japan

(72) Inventor
Yasukuni Nonaka

(74) Agent and/or Address for Service
Haseltine Lake & Co
Hazlitt House, 28 Southampton Buildings,
Chancery Lane, London, WC2A 1AT, United Kingdom

(51) INT CL⁴
H01Q 1/12

(52) UK CL (Edition J)
H1Q QKN
U1S S2212

(56) Documents cited
GB 0800306 A   EP 0114543 A2

(58) Field of search
UK CL (Edition J) H1Q QKE QKH QKN
INT CL⁴ H01Q

(54) Portable antenna apparatus for satellite communication

(57) A portable satellite broadcast signal antenna apparatus for manually transporting a satellite signal converter. The apparatus includes an antenna device (100a, 100b) for receiving the satellite broadcast signals, a device (208) for adjusting the angle of the antenna device (100a, 100b) with respect to a predetermined plane for aligning the antenna device (100a, 100b) with a satellite, and a housing (202, 204) defining an interior space for receiving the satellite signal converter and the angle adjusting device (208) therein and having an exterior surface bearing the antenna device (100a, 100b).

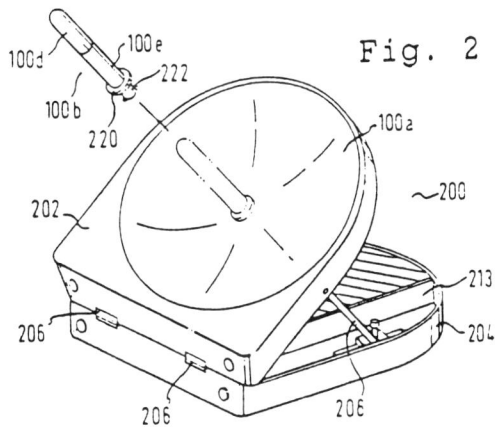

Fig. 2

GB 2 208 189 A

Front page of a published British patent application

(12) **UK Patent** (19) **GB** (11) **2 178 737** (13) **B**

(54) Title of Invention

Process for fabricating optical fibre

(51) INT CL⁴ ; **C03B 37/027**

(21) Application No
8615513.2

(22) Date of filing
25 Jun 1986

(30) Priority data

(31) **60138754**
60148067

(32) **25 Jun 1985**
5 Jul 1985

(33) **JP**

(43) Application published
18 Feb 1987

(45) Patent published
15 Mar 1989

(52) Domestic classification
(Edition J)
C1M MBD M401

(56) Documents cited
GB 1427327 A
GB 1559097 A

(58) Field of search

As for published application
2178737 A *viz:*
UK CL **C1M**
INT CL⁴ **C03B**
updated as appropriate

(73) Proprietor(s)
**The Furukawa Electric
Co Ltd**

(Incorporated in Japan)

No 6-1 Marunouchi 2-chome
Chiyoda-ku
**Tokyo 100
Japan**

(72) Inventor(s)
**Tamotsu Kamyia
Yasuhiro Shibayama
Nobuhito Uchiyama
Noboru Sato**

(74) Agent and/or
Address for Service
**Urquhart-Dykes & Lord
Midsummer House
419B Midsummer Boulevard
Central Milton Keynes
MK9 3BN**

Front page of a British granted patent

After the front page come the illustrations, introductory material, the description of the invention, and on the final pages the claims.

Page 12 shows the front page of a granted, or 'B' specification. The arrangement of the information is somewhat different from that on the 'A' document, and no abstract or drawing is given. The claims in the 'B' document are the definitive monopoly given to the applicant (unless opposition is successful).

Abstracts

Abstracts of published British patent specifications are available from the first patent in 1617. They are arranged in four series, which all consist of subject volumes (which contain indexes by subject and by name) with the abstracts in numerical order in each volume. An exception is 1979-80, when the abstracts were arranged within the subject volumes in order of the headings ('subclasses').

Before the 1977 Patent Act the summaries, called abridgments, were prepared by Patent Office examiners. Since then the full front pages, including abstracts prepared by the applicants, are supplied. The Patent Office still has the power to amend the abstracts before publication.

The most recent series dates from 1962 onwards. Each week the front pages of published patent applications, arranged numerically within British classification divisions, are issued. Published volumes are substituted for them a couple of years later. These volumes contain cumulated annual subject-matter indexes.

An abstract for a particular specification can be identified by using the *Divisional Allotment Index* (formerly the *Group Allotment Index*). These numerical listings refer to the relevant subject volume. They commence with British patent 340,001, published in 1931.

It must be remembered that the increasing number of European Patent Convention and Patent Cooperation Treaty specifications since 1978 means that making a search through these British indexes is no longer comprehensive. This is because the other published applications have as much potential relevance for prior art and infringement purposes.

The gazette

The *Official Journal (Patents)*, is an alerting journal which is widely used for current awareness purposes. Its format changed somewhat during early 1989. The illustrations used reflect the new format.

The *OJ(P)* first announces patent applications about six weeks after their submission to the Patent Office. Page 14 gives an example of the information given. The first Citizen Watch Company application, for example, had the title 'Scrap collector system'. It was given the application number 8903498.7 when it was 'lodged' or applied for on 16 February 1989. It was originally applied for in Japan (JP) on 27 February 1988, where it was given the application number 63019601. The advantages of

5 April 1989   Official Journal (Patents)

Applications for Patents—cont

Cadbury Limited Confectionery box
Date Lodged: 18 Feb 1989
GB8903732.9

—Confectionery box
Date Lodged: 18 Feb 1989
GB8903733.7

Cain, Paul P and Nye, Philip P
The valve
Date Lodged: 20 Feb 1989
GB8903769.1

Caldwell, Kenneth Improvements to
inflated cushions and mattresses
Date Lodged: 17 Feb 1989
GB8903613.1

Caligen Foam Limited Cleaning and
scouring products
Date Lodged: 14 Feb 1989
GB8903247.8

—Cleaning and scouring products
Date Lodged: 14 Feb 1989
GB8903248.6

Callaghan, Leo C Drinker for a
domestic animal
Date Lodged: 15 Feb 1989
Priorities: [GB8822494 24 Sep 1988]
GB8903475.5

Cambridge Consultants Limited
Coherent tracking sensor
Date Lodged: 18 Feb 1989
GB8903725.3

Cambridge Endeavour Limited
Detection or investigation of proteins
Date Lodged: 23 Feb 1989
GB8904129.7

Cann, William A H Window opener
and stay
Date Lodged: 18 Feb 1989
GB8903716.2

Cannon Industries Limited Flue
systems
Date Lodged: 22 Feb 1989
GB8904013.3

Cannon Jewellery Limited Jewellery
Date Lodged: 23 Feb 1989
Priorities: [GB8805216 04 Mar 1988]
GB8904197.4

Canon Kabushiki Kaisha Data
recording apparatus
Date Lodged: 14 Feb 1989
[24 Feb 1986]
Priorities: [JP60035612 25 Feb 1985]
[JP60047249 08 Mar 1985]
[JP60045957 07 Mar 1985]
GB8903310.4

—Information processing apparatus
Date Lodged: 16 Feb 1989
Priorities: [JP63032653 17 Feb 1988]
GB8903598.4

Carello Lighting Plc Lamp reflector
Date Lodged: 18 Feb 1989
GB8903738.6

Carey, Maurice Electronic darts
scoreboard (for supply to bars. purpose:
advertising)
Date Lodged: 22 Feb 1989
GB8903954.9

—Roller wire brush
Date Lodged: 22 Feb 1989
GB8904053.9

—Star and planet finder
Date Lodged: 22 Feb 1989
GB8904056.2

**Cebria Navarro, Jose L and Cebria
Navarro, Victor M** Anti-theft device
for motor vehicles
Date Lodged: 17 Feb 1989
Priorities: [ES00473 18 Feb 1988]
GB8903680.0

Cebria Navarro, Victor M *See* Cebria
Navarro, Jose L

Celltech Limited Vector
Date Lodged: 22 Feb 1989
GB8904009.1

Centa, Antony R Access covers for
manholes
Date Lodged: 17 Feb 1989
Priorities: [GB8804652 27 Feb 1988]
GB8903714.7

—Floor divider strips
Date Lodged: 17 Feb 1989
GB8903715.4

Central Glass Company Limited
Vehicle window glass antenna suited to
reception of FM radio and TV
broadcasting
Date Lodged: 21 Feb 1989
Priorities: [JP63040723 25 Feb 1988]
[JP63107787 28 Apr 1988]
GB8903896.2

**Central Glass Company Limited and
Ikeda Glass Industrial Co Ltd**
Method of fitting plate member with
supportive or protective member of
moulded resin
Date Lodged: 20 Feb 1989
Priorities: [JP6335700 18 Feb 1988]
GB8903788.1

Cha, Eung H Contoured car wash brush
Date Lodged: 03 Jan 1989
GB8903934.1

**Chapelhow, Stephen L and Gooding,
Stephen W and Nolan, Dermot W**
Apparatus for playing a game
Date Lodged: 17 Feb 1989
GB8903702.2

Charlton, John R Shock absorber kit for
wheel castors
Date Lodged: 14 Feb 1989
GB8903356.7

—Portable sprung seat for wheelchairs
Date Lodged: 17 Feb 1989
GB8903612.3

**Chick, Graeme D and Hughes,
Michael A J** Plug tidy and storage unit
Date Lodged: 22 Feb 1989
GB8904073.7

**Chick, Graeme D and Hughes,
Michael A J** The paint brush tip
Date Lodged: 22 Feb 1989
GB8904074.5

Chiu, Su C Built up conveyor
Date Lodged: 15 Feb 1989
GB8903460.7

Chute, Nigel A Multi-part kitchen tidy
Date Lodged: 17 Feb 1989
GB8903614.9

—Folding, freestanding screen
Date Lodged: 17 Feb 1989
GB8903615.6

Ciba-Geigy AG Polymerisation of
maleic anhydride
Date Lodged: 14 Feb 1989
GB8903330.2

—Azo dyes, processes for their
preparation and their use
Date Lodged: 22 Feb 1989
Priorities: [CH729 26 Feb 1988]
[CH1414 18 Apr 1988]
GB8904045.5

Citizen Watch Co Ltd Scrap collector
system
Date Lodged: 16 Feb 1989
Priorities: [JP63024488 27 Feb 1988]
GB8903498.7

—IC card
Date Lodged: 15 Feb 1989
Priorities: [JP63019061 16 Feb 1988]
GB8903465.6

Coal Industry (Patents) Limited
Improvements relating to arch setting
Date Lodged: 16 Feb 1989
GB8903543.0

Coates-Smith, Ronald Water vapour
barrier
Date Lodged: 14 Feb 1989
GB8903235.3

Cochrane, Ian R Clamp
Date Lodged: 21 Feb 1989
GB8903883.0

Colledge, Cecil A combination back
pack, camp bed and frame tent
Date Lodged: 21 Feb 1989
GB8903858.2

Collins, John Plastic theft strip
Date Lodged: 20 Feb 1989
GB8903760.0

Conway Limited Security co-axial cable
interface transmitter
Date Lodged: 14 Feb 1989
GB8903344.3

Cookson Graphics Plc Lithographic
printing plate
Date Lodged: 17 Feb 1989
GB8903691.7

Cookson Group Plc Magnetic
compositions
Date Lodged: 14 Feb 1989
GB8903329.4

Coopers Animal Health Limited
Chemical compositions
Date Lodged: 22 Feb 1989
GB8903978.8

Coppernob Limited Improvements
relating to articles of clothing
Date Lodged: 21 Feb 1989
GB8903859.0

Cork Amusement Centre Limited
Improvements in or relating to a credit-
freed machine
Date Lodged: 14 Feb 1989
GB8903307.0

Corrigan, Nigel M Radio
communications
Date Lodged: 14 Feb 1989
GB8903342.7

Cotutsca, Peter Display devices
Date Lodged: 14 Feb 1989
GB8903318.7

Courtaulds Plc Joining coated fabric
Date Lodged: 14 Feb 1989
Priorities: [GB8803456 15 Feb 1988]
GB8903291.6

—Flexible fabric thermal insulators
Date Lodged: 17 Feb 1989
GB8903641.2

Name index to British patent filings

being given the priority information are that it indicates the provisional novelty date, and suggests that publication will occur up to a year earlier than domestic priority applications, i.e. 18 months after the earliest date supplied.

When the applications are published, brief details are given in the *OJ(P)* in British abbreviated classified order. An example is given on page 16. Since the applications are arranged in classified order before they are numbered, this forms a numerical sequence as well. References are given from subordinate to the more important subject aspects of a patent application. For example, there is a reference from F4S to F4T concerning GB 2208426.

The information given for GB 2208435 (at G1B on page 16) is as follows. The filing number has after it the date of filing of the present application, 1 August 1988. The applicant's name is given, and the title, 'Sampling'. Any priority details are given. These are often foreign, but sometimes as here refer to earlier attempts in Britain to file the application. Although the applicant has resubmitted the application, the original filing date is still the provisional novelty date. The British classification headings are given again followed by the 'Int C^4' numbers. These are the classification numbers of the 4th edition of the International Patent Classification.

The published applications are indexed by applicant in each issue of the *OJ(P)*, as shown on page 17. There are references from the inventors to their companies, as for Thomas Mintel, and for each firm the inventors' names are given in brackets. These indexes of applicants are cumulated annually into published volumes.

Similar bibliographic details are given for granted patents which are indexed by subject and by applicant. There are no cumulative indexes for granted patents.

European Patent Convention information in the *OJ(P)*

Besides providing information on British patents published through the Patent Office, the *OJ(P)* also covers those European Patent Convention patents that designate Britain, since they are also national patents. They are called 'European Patents (UK)'.

This information consists of listings of granted patents, with applicant but not title, and 'Non-English' if the patent is in a foreign language; listings of translations of European granted patents, and of claims of applications; European patents treated as void due to failure to file translations; revoked patents; and European patents that have ceased in Britain through non payment of renewal fees.

British and European patents

Some British applicants routinely abandon a British application and continue with the European Patent Convention (EP) equivalent. Others

30 March 1989   Official Journal (Patents)   883

Applications published under Section 16(1)—cont

F2H

GB2208417 (GB8818359.5) 02 May 1988
[02 Aug 1988]
GIANNUZZI, LOUIS N
Roof anchor and stress plate assembly
Priorities: [US081016 03 Aug 1987]
[US215307 02 Jul 1988]
UKC Headings: F2H Int Cl⁴ F16B 39/32

F2S

GB2208418 (GB8809988.2) 27 Apr 1988
SHIMIZU CONSTRUCTION CO
LTD (INCORPORATED IN JAPAN)
Restraining vibration of a floor
Priorities: [JP62103771 27 Apr 1987]
[JP62103770 27 Apr 1987]
UKC Headings: F2S U1S Int Cl⁴ F16F
7/10

GB2208419 (GB8809989.0) 27 Apr 1988
SHIMIZU CONSTRUCTION CO
LTD (INCORPORATED IN JAPAN)
Restraining vibration of a structure
Priorities: [JP6264951 28 Apr 1987]
[JP62105571 28 Apr 1987]
UKC Headings: F2S U1S Int Cl⁴ F16F
7/10

GB2208420 (GB8818313.2) 02 Aug 1988
POLIPREN S R L (INCORPORATED
IN ITALY)
A resiliently deformable element for use
as a buffer stop in a motor vehicle
suspension
Priorities: [IT53593U 07 Aug 1987]
UKC Headings: F2S U1S Int Cl⁴ F16F
9/32

F2V

GB2208421 (GB8718080.8) 30 Jul 1987
KENT PROCESS CONTROL
LIMITED (INCORPORATED IN
UNITED KINGDOM)
Flow control valve
UKC Headings: F2V Int Cl⁴ F16K 1/54
F16K 47/00

F4G

GB2208422 (GB8818151.6) 29 Jul 1988
HOTPOINT LIMITED
(INCORPORATED IN UNITED
KINGDOM)
Venting systems
Priorities: [GB8718176 31 Jul 1987]
UKC Headings: F4G F2G U1S Int Cl⁴
F16L 41/08

F4K

GB2208423 (GB8718500.5) 05 Aug 1987
STORDY COMBUSTION
ENGINEERING LIMITED
(INCORPORATED IN UNITED
KINGDOM)
Furnace burners with regenerative heat
exchangers
UKC Headings: F4K F4T U1S Int Cl⁴
F28D 17/00 F23L 15/02 F28D 19/00

F4R

GB2208424 (GB8806785.5) 22 Mar 1988
ESKOFOT A/S (INCORPORATED IN
DENMARK)
Area exposure apparatus
Priorities: [DK152387 25 Mar 1987]
UKC Headings: F4R U1S Int Cl⁴ F21V
7/12

GB2208425 (GB8818754.7) 08 Aug 1988
FAIRFIELD, GEOFFREY D
Display lighting apparatus
Priorities: [GB8718661 05 Aug 1987]
UKC Headings: F4R U1S Int Cl⁴ F21V
21/34

F4S

GB2208426 *See* entry under Heading F4T

F4T

GB2208423 *See* entry under Heading F4K

GB2208426 (GB8814796.2) 22 Jun 1988
J EBERSPACHER (INCORPORATED
IN FR GERMANY)
Apparatus for preheating liquid fuel for
heating apparatus
Priorities: [DE3721834 02 Jul 1987]
UKC Headings: F4T F4S U1S Int Cl⁴
F23D 11/44

F4V

GB2208427 (GB8718342.2) 03 Aug 1987
DUFAYLITE DEVELOPMENTS
LIMITED (INCORPORATED IN
UNITED KINGDOM)
Smoke diverter valve in air duct
UKC Headings: F4V A5A Int Cl⁴ F24F
13/14 F16K 1/20

F4W

GB2208428 (GB8718848.8) 08 Aug 1987
HEPWORTH HEATING LIMITED
(INCORPORATED IN UNITED
KINGDOM)
Space heating appliances
UKC Headings: F4W Int Cl⁴ F24C 15/06

GB2208429 (GB8818254.8) 01 Aug 1988
KOPONEN, TAUNO K
KOPONEN, LAILA T
Convection air heating guard
Priorities: [FI873332 31 Jul 1987]
UKC Headings: F4W Int Cl⁴ F24C 15/36

G1A

GB2208430 (GB8717542.8) 24 Jul 1987
HILGER ANALYTICAL LIMITED
(INCORPORATED IN UNITED
KINGDOM)
Monochromators
UKC Headings: G1A Int Cl⁴ G01J 3/06
G01J 3/18

GB2208431 (GB8817731.6) 26 Jul 1988
LINK ANALYTICAL LIMITED
(INCORPORATED IN UNITED
KINGDOM)
Conditioning X-ray detectors
Priorities: [GB8718531 05 Aug 1987]
UKC Headings: G1A H1D U1S Int Cl⁴
H01J 37/244 G01T 7/00

GB2208432 (GB8817877.7) 27 Jul 1988
JEOL LIMITED (INCORPORATED
IN JAPAN)
Charged particle beam analyser
Priorities: [JP62192098 31 Jul 1987]
UKC Headings: G1A H4F Int Cl⁴ G01N
23/225

GB2208433 (GB8818111.0) 29 Jul 1988
VEGLIA BORLETTI S R L
(INCORPORATED IN ITALY)
Detecting water droplets on a vehicle
window and controlling windscreen
wiper in response thereto
Priorities: [IT5356887U 31 Jul 1987]
UKC Headings: G1A U1S Int Cl⁴ G01N
21/84

GB2208434 (GB8818112.8) 29 Jul 1988
VEGLIA BORLETTI S R L
(INCORPORATED IN ITALY)
Detecting water droplets on a vehicle
window and controlling windscreen
wiper in response thereto
Priorities: [IT535698U 31 Jul 1987]
UKC Headings: G1A U1S Int Cl⁴ G01N
21/84

G1B

GB2208435 (GB8818272.0) 01 Aug 1988
NATIONAL RESEARCH
DEVELOPMENT CORPORATION
(INCORPORATED IN UNITED
KINGDOM)
Sampling
Priorities: [GB8718232 31 Jul 1987]
UKC Headings: G1B U1S Int Cl⁴ G01N
1/02 C12Q 1/24

Classified entries for published British applications

Name Index of Applications Published—cont

Itoh, Toshiaki *See* Central Glass Company Limited (Incorporated in Japan)

Izawa, Masataka *See* Honda Giken Kogyo Kabushiki Kaisha (Incorporated in Japan)

J Eberspacher (Incorporated in FR Germany) (Reiser, Peter) (Schodt, Adolf) (Mohring, Fritz)
F4S F4T U1S GB2208426

Jacquelin, Patrice *See* Matra (Incorporated in France)

Jasper, Hans *See* Stordy Combustion Engineering Limited (Incorporated in United Kingdom)

Jeol Limited (Incorporated in Japan) (Ono, Akishige)
G1A H4F GB2208432

Jervis, Audrey P *See* Jervis, Hubert B

Jervis, Hubert B (Jervis, Audrey P)
A4M B7A U1S GB2208353

John Wyeth & Brother Limited (Incorporated in United Kingdom) (Cliffe, Ian A)
A5B C2C U1S GB2208385

Johnson, Martin S *See* IMI Cornelius (UK) Limited (Incorporated in United Kingdom)

Jowett, Peter
B1E U1S GB2208357

Joyce, Peter W *See* Dufaylite Developments Limited (Incorporated in United Kingdom)

Kanzaki, Hisao *See* Fuji Photo Film Co Ltd (Incorporated in Japan)

Karasek, Mark L *See* Wilson Sporting Goods Co (Incorporated in USA-Delaware)

Kasa, Koichi *See* Pioneer Electronic Corporation (Incorporated in Japan)

Kawamoto, Yoshimichi *See* Honda Giken Kogyo Kabushiki Kaisha (Incorporated in Japan)

Kent Process Control Limited (Incorporated in United Kingdom) (Gates, Eric)
F2V GB2208421

Kiebert GmbH & Co Kommanditgesellschaft (Incorporated in FR Germany) (Kleefeldt, Frank)
E2A U1S GB2208407

Kishimoto, Takashi *See* Matsushita Electric Works Limited (Incorporated in Japan)

Kleefeldt, Frank *See* Kiekert GmbH & Co Kommanditgesellschaft (Incorporated in FR Germany)

Kobayashi, Seishiro *See* Anritsu Meter Co Ltd (Incorporated in Japan)

Kohata, Takaski *See* Honda Giken Kogyo Kabushiki Kaisha (Incorporated in Japan)

Koponen, Laila T *See* Koponen, Tauno K

Koponen, Tauno K (Koponen, Laila T)
F4W GB2208429

Lazare, Francois *See* Irete SA (Incorporated in Switzerland)

Lebedev, Jury A *See* Gorno-Altaisky Gosudarstvenny Pedagogichesky Institut (Incorporated in Soviet Union)

Licinvest AG (Incorporated in Switzerland) (Ackeret, Peter)
B6E U1S GB2208371

Link Analytical Limited (Incorporated in United Kingdom) (Lowe, Barrie G) (Tyrrell, Stuart G J)
G1A H1D U1S GB2208431

Lowe, Barrie G *See* Link Analytical Limited (Incorporated in United Kingdom)

ML Engineering (Plymouth) Limited (Incorporated in United Kingdom) (Bartram, Christopher)
G4Q GB2208449

M-B-W Inc (Incorporated in USA-Wisconsin) (Artzberger, Thomas G)
E1G GB2208401

Makita, Kensuke *See* Central Glass Company Limited (Incorporated in Japan)

Man Design Co Ltd (Incorporated in Japan) (Wakatsuki, Yoshio) (Shimizu, Giichiro) (Okuyama, Toshiharu)
G1N U1S GB2208440

Mander, Frederick W
A4G GB2208350 A4G GB2208351 A4G B8P U1S GB2208352

Marconi Electronic Devices Limited (Incorporated in United Kingdom) (Mitchell, Alastair)
H1M GB2208453

Martin, William J *See* National Research Development Corporation (Incorporated in United Kingdom)

Masternet Limited (Incorporated in United Kingdom) (Smith, Stephen J)
B8P U1S GB2208380

Matra (Incorporated in France) (Perez, Ellio) (Deneville, Pierre) (Jacquelin, Patrice)
H2A H2F U1S GB2208455

Matsuguchi, Tadashi *See* Daimatsu Kagaku Kogyo Co Limited (Incorporated in Japan)

Matsuguchi, Noboru *See* Daimatsu Kagaku Kogyo Co Limited (Incorporated in Japan)

Matsuguchi, Noboru *See* Daimatsu Kagaku Kogyo Co Limited (Incorporated in Japan)

Matsuguchi, Tadashi *See* Daimatsu Kagaku Kogyo Co Limited (Incorporated in Japan)

Matsuki, Taketo *See* Sumitomo Rubber Industries Ltd (Incorporated in Japan)

Matsumoto, Shusaku *See* Dai-Ichi Kogyo Seiyaku Co Ltd (Incorporated in Japan)

Matsuoka, Hiroshi *See* Mitsubishi Denki Kabushiki Kaisha (Incorporated in Japan)

Matsushita Electric Works Limited (Incorporated in Japan) (Kishomoto, Takashi) (Hirao, Shyozo) (Yokokawa, Hiroshi) (Takahama, Koichi) (Yokoyama, Mazaru)
C1A GB2208382

Matsushita, Hiroomi *See* Sumitomo Rubber Industries Ltd (Incorporated in Japan)

Mattinson, Beverley I *See* Industrial Gloves (Speke) Limited (Incorporated in United Kingdom)

Matuda, Naohiko *See* Central Glass Company Limited (Incorporated in Japan)

Mengel, William H *See* RCA Licensing Corporation (Incorporated in USA-New Jersey)

Mevissen, Peter *See* Monforts, A GmbH & Co (Incorporated in FR Germany)

Mintel, Thomas E *See* Colgate-Palmolive Company (Incorporated in USA-Delaware)

Misevich, Kenneth W *See* Colgate-Palmolive Company (Incorporated in USA-Delaware)

Mitchell, Alistair *See* Marconi Electronic Devices Limited (Incorporated in United Kingdom)

Mitsubishi Denki Kabushiki Kaisha (Incorporated in Japan) (Matsuoka, Hiroshi)
H1K U1S GB2208452

Mitsubishi Kasei Corporation (Incorporated in Japan) (Sagara, Kazuo) (Nakayama, Mamoru) (Yamada, Yasuo)
B8A B8S GB2208378

Miyao, Kouji *See* Sharp Kabushiki Kaisha (Incorporated in Japan)

Mohring, Fritz *See* Eberspacher, J (Incorporated in FR Germany)

Name index to published British applications

17

allow both applications to run through to grant. However, if both are granted the Patent Office will insist on the British patent being revoked.

Case law

British patent court cases can be divided into two main kinds: Patent Office Court decisions, and High Court decisions.

Patent Office decisions are either hearings when an application for an industrial property right has been refused, or are used as a cheap way of settling disputes (often on licencing). They cannot be cited as precedents.

On appeal, cases from both the Patent Office and the new County Patent Court can go on to the High Court. Cases can also begin in the High Court.

Formally published decisions serials and indexes are listed on page 106.

The Patent Office's patent information services

Besides examining and granting patents, the British Patent Office offers both publications and services. Many of these are listed (with current prices) on the inside back page of each issue of the *Official Journal (Patents)*. The last issue of every month gives many details of services provided by the Patent Office.

These services include supplying copies of British patents, and translations of European granted patents. There is also a current awareness service whereby either lists of patent numbers, or actual specifications, within British classification profiles can be provided. These services are dealt with by:

Sale Branch, The Patent Office, Unit 6, Nine Mile Point, Cwmfelinfach, Newport, Gwent NP1 7HZ. Tel: 0633 246151. Fax: 0495 200915.

OPTICS, the Patent Office automation system for recording data relating to patents published after the 1977 Act (with some earlier material), is still being developed, and is not yet fully available to the public. However, it is possible to obtain legal information such as confirmation of the status of a patent; to whom a patent is being licensed; and to obtain copies of the 'shell' or file of correspondence relating to a published patent application. This includes a more detailed search report than that published in the application, and is similar to the European Patent Office search reports described in Chapter 3. The Patent Office can be contacted on 071 438 4724 for details on how to use these services.

The Patent Office also contains the Search and Advisory Service. This offers a comprehensive range of online patents databases. It also has access to two hundred specialist patent examiners who can offer expert search services, and a large classified set of patent documents. Their address is:

Search and Advisory Service, Patent Office, Hazlitt House, 45 Southampton Buildings, Chancery Lane, London WC2A 1AR. Tel: 071 438 4747/48. Fax: 071 438 4750.

SRIS and the British patents

This section explains special practices at SRIS in making British patent information available.

The British applications published each week are numerically arranged on a table before being replaced by the following week's publications. Since the applications are published in British classified order it is easy to identify specifications on a particular topic.

The granted patents for the last three weeks are kept numerically in separate boxes for each week before shelving in the main sequence.

SRIS keeps recent British abstracts in boxes, arranged by subject group, for each year. Subject access is provided by a separate series of black binders containing weekly updates, each in classified sequence, of the specifications published that week.

There is a telephone at the British Patents Desk that can be used to talk to Patent Office examiners about how to classify a particular subject by British classification. This service is free.

The filed applications cannot be searched by subject before they are published, other than online using the title information. However, searches by applicant can be carried out in annual cumulations, which are kept on card-indexes at SRIS (the current year plus the two previous years are on open access). Page 20 shows three cards. The information is exactly the same as in the *OJ(P)*. These cards are filed in the current cumulation about two weeks before the same information appears in the *OJ(P)*.

The *OJ(P)* is useful for current awareness but it can also be important to find the published numbers for an application that was noticed years earlier in the *OJ(P)*, or to find out what subsequently happened to a published application.

SRIS maintains two sets of nonofficial registers, which are compiled by using information given in the *OJ(P)*. They are updated as soon as the journals are published.

The first of these is the Application Register, a page of which is shown on page 21. This gives the published number for every application number or, if it was not published, the outcome of the application. Some were withdrawn before publication. Others needed to be submitted more than once, and a stamp is given showing earlier or later applications, as in 8611857. In other cases it has been possible, through routine scanning of various journals, to give foreign equivalents, as in the case of 8611863, where EP 220894 and US 4811616 give essentially the same technical information.

Information on translations of European Patent Convention patents, and on void patents, is entered in SRIS's non-official EPO Publication Register (see page 35).

The second set of registers is the Stages of Progress Register. An example of its contents is given on page 23. The first column gives the date of publication for the applications, which in this case was 14 January 1981. Some were subsequently withdrawn, and some were 'no case', where a number was incorrectly assigned. The double column '2nd publication'

```
8904200.6  23 FEB 89 |
                      |----------------
Baillie,David John    |
Anning,Michael John Phillip |
                      |
Improvements in or relating to a motor vehicle accessory |----------------
                      |
No of Priorities:  0  Earliest priority: |----------------
```

```
8916789.4  21 JUL 89 |
                      |----------------
Ansaldo S p A         |
                      |
                      |
Heat exchanger tube array,and manufacturing process |----------------
                      |
No of Priorities:  1  Earliest priority: IT12526      22 JUL 1988 |----------------
```

```
8918093.9  08 AUG 89 |
                      |----------------
Anson Limited         |
                      |
                      |
Improved pipeline couplings |----------------
                      |
No of Priorities:  0  Earliest priority: |----------------
```

Card index to British patent filings

APPLICATION NUMBER	DOCUMENT NUMBER	REMARKS
8611851	2182204 WITHDRAWN	
2	SEE OJ (P)	—
3	WITHDRAWN SEE (P)	EP 249346
4	2177173	
5	2175460	
6	WITHDRAWN SEE	
7	2175456	Pr. APP. 12625/1985 SEE (P) 25.6.86
8	2175361 WITHDRAWN	SEE OJ (P) 14.2.90
9	2175997	
8611860	2175923	
1	2190493	
2	2177798	
3	2181912	EP 220 894 US 4811616
4	2175436	Pr. APP 12346/1985 SEE OJ(P) 25.6.86
5	WITHDRAWN 18/11/87 SEE OJ (P)	EP 246087 US 4822-348
2.12.87 WITHDRAWN SEE OJ(P) 6	2175184	Pr. APP. 12259/1985 SEE OJ(P) 25/6/86 20144/1935 REMARKS 25/6/86 ✳
7	WITHDRAWN	
8	WITHDRAWN SEE OJ(P)	
9	SEE OJ (P)	
8611870	WITHDRAWN SEE OJ (P)	LATER APP 8711527 SEE OJ(P) 17.6.87
1	2175268	Pr. APP. 12621/1985 SEE OJ(P) 25.6.86
2	2175188 WITHDRAWN	Pr. APP. 12559/1985 SEE OJ(P) 25.6.86
3	SEE OJ (P) 3-9-87	LATER APP 8711438 SEE OJ(P) 17/6/87
4	WITHDRAWN SEE OJ(P) 3-9-87	LATER APP 8711439 SEE OJ(P) 17/6/87 EP 249327
5	2177259	
6	WITHDRAWN SEE OJ(P)	
7	2176016	APP. 13586/1985 SEE OJ(P) 25.6.86
8	2175569	APP. 13587/1985 SEE OJ(P) 25.6.86
9	OJ (P)	

Non-official Application Register

and 'Grant' gives a single date, since these are now the same. Under the 1949 Act patents were only granted when they were 'sealed', between four and seven months after publication.

The lapsed column is for patents which lapse due to the nonpayment of renewal fees, which are due annually from five years after application. It is possible to restore apparently lapsed patents under certain conditions so this information should be treated with caution. The last column shows the date when patents that ran their full term expired.

SRIS holds transcripts of many industrial property decisions given after hearings in the British courts from 1970 onwards. Many of them are never formally published. They are the Patent Office and High Court series, and in the new Patents County Court series, which began in 1990.

These series ought to be complete except for the High Court series, of which a selection only is held.

Card indexes of the parties involved in all three series from 1976 are available at SRIS.

The licence of right provision

The 'Remarks' column can contain many different comments on status changes. One of these is a red 'L of R' or if in an old register an asterisk. This indicates that the owner of the patent has declared a licence of right, by which in exchange for only paying half the renewal fees, the invention is licenced to any interested party on a non-exclusive basis. A licence of right can also be imposed on the owner of patents if the patent is not 'worked'.

Further reading

The readability and information content of new law abstracts and old law abridgments, L. Blakeborough and C. Oppenheim. *CIPA*, Dec 1980. 10 (3), 86-92.

Patent litigation in the UK: the new Patents County Court. A. Webb. *European Intellectual Property Review*, June 1991, 13 (6), 203-212 [a useful review of litigation procedure].

DOCUMENT NUMBER OF 1st PUBLICATION	1st PUBLICATION	WITHDRAWN	DATE OF OJ(P) ANNOUNCING: 2nd PUBLICATION	GRANT	LAPSING	EXPIRY	REMARKS
2064800	17 JUN 1981		[5EC 24 ...] (handwritten)				
1				15 [...] 1983	27 JUN 1990		
2				2 MAR 1983	25 JUL 1990		
3				1 SEP 1983 / NOV 9 1986			
4				7 DEC 1983			
5				[illegible]			
6				10 MAY 1984			
7				12 OCT 1983	24 APR 1991		
8		9 SEP 1984					
9				23 MAY 1984	09 JUL 1986		
2064810				15 SEP 1984			
1				29 JUN 1983			
2		[...] 1984					
3				1 SEP 1983	15JUL87		
4				29 JUN 1983	27 JUL 88		
5				10 MAY 1984			
6		22 AUG 1984		3 AUG [...]	6 JUL 1968		
7				23 NOV 1983			
8				25 JUL 1984			
9				18 JAN 1984			
2064820				2.2.83	13 JUN 1990		
1				[...] 1983	14 JUN 1989		
2				22 JUN 1983	09 JUL 1986		
3				30 MAR 1983	10 JUL 1987		
4				2 JUN 1983	10 JUL 1987		
5							
6		[...]					

Non-official Register of Stages of Progress

This page intentionally left blank

3 EUROPEAN PATENT CONVENTION PUBLICATIONS

The European Patent Convention, under which applications for patents have been made since 1978, enables patents to be obtained in up to 14 designated European countries following a single application which may be submitted in English, French or German. It is administered by the European Patent Office (EPO) at Munich.

At present over half of the applications are published in English. The member countries of the Convention, and their country codes as shown on the specifications, are, as of the 1 January 1992:

Austria	(AT)	Luxembourg	(LU)
Belgium	(BE)	Monaco	(MC)
Denmark	(DK)	Netherlands	(NL)
France	(FR)	Portugal	(PT)
Germany	(DE)	Spain	(ES)
Greece	(GR)	Sweden	(SE)
Italy	(IT)	Switzerland	(CH)
Liechtenstein	(LI)	United Kingdom	(GB)

Ireland (IE) is hoping to join during 1992.

Numeration

Applications are given a number such as 81304643 when they are filed at the EPO. The first two digits indicate the year of filing, the third digit indicates the filing office among the member countries and the remaining five digits form a sequence that begins from 1 each year. The numbering of published applications is continuous, beginning at EP 0000001.

Patenting procedure

The patenting procedure is similar to that under the British 1977 Act in that the 'A' specification is published 18 months after the priority date. Examination is on request within six months of publication of the search report, and, if accepted, the specification is published a second time ('B' specification). The procedure is illustrated on page 27.

The first applications were filed on 1 June 1978, the first specifications were published in December 1978 and the first patents granted in April 1980. Over 60,000 applications were published in 1990.

'A1' and 'A2' specifications are published weekly (18 months from the priority date) in numerical order, generally on Wednesdays. If an application for a European Patent is made by first applying under the Patent Cooperation Treaty (PCT) (see page 37) the specification will not be republished by the European Patent Office although it will be given a number.

An exception to this rule is made if the PCT application was not published in English, French or German. To inspect the specification for a 'Euro-PCT' application the sequence of PCT documents must be sought.

The granted patents ('B' specification) are also published weekly in the original language but the claims are in English, French and German. It should contain no technical disclosure which was not in the 'A' specification but is important in that it defines the scope of the protection given by the patent. The front page of the second publication has a summary of the search report but no abstract or drawing.

Opposition to granted patents can be initiated within nine months of grant. The opposition is then dealt with by the EPO.

Opposition taking place after the nine month period must be pursued separately through the national courts of the designated countries.

Renewal fees are paid to the individual patent offices, since the patents are then regarded as separate national patents. The patent can be allowed to lapse in some countries while it continues to be protected in others. The maximum term of protection is 20 years from the filing date at the European Patent Office.

Specifications

The front page of the first published 'A' specification contains bibliographic information, including the designated states, an abstract and a significant drawing where appropriate. The search report may be published with the specification or separately later.

The 'kind of document code' which follows the number has the following significance:

A1	First publication of the specification with search report.
A2	First publication of the specification without search report.
A3	Later publication of search report with revised front page of specification.
B1	Second publication of the specification, the granted patent.
B2	Amended granted patent.

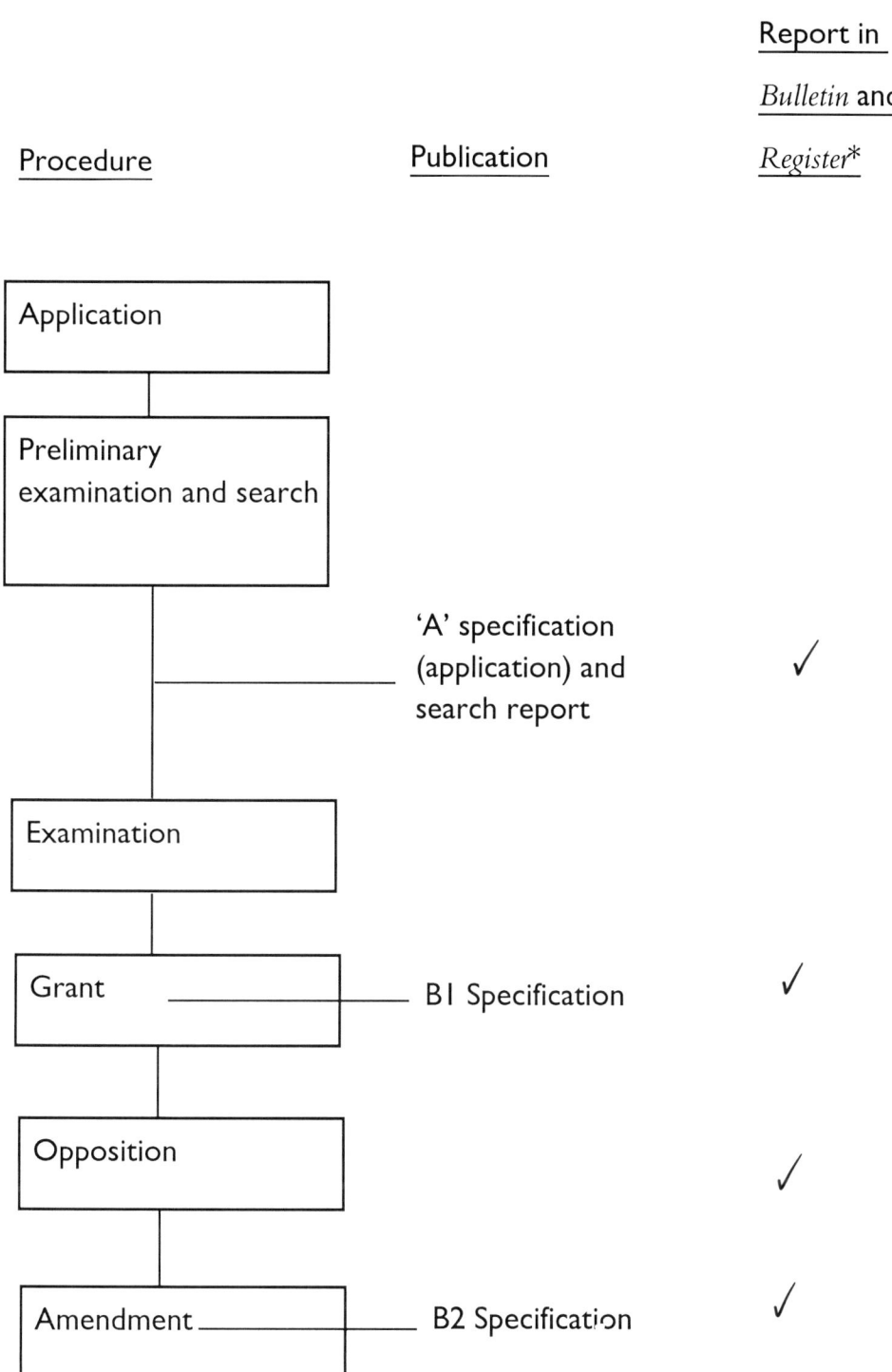

Procedure
Publication
<u>Report in</u>
Bulletin and
*Register**

Application

Preliminary examination and search

'A' specification (application) and search report ✓

Examination

Grant —————— B1 Specification ✓

Opposition ✓

Amendment —————— B2 Specification ✓

⋆ *European Patent Bulletin* and *European Register*

Progress of a European patent application

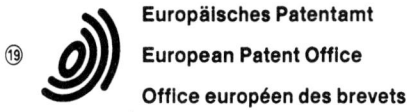

(19) **Europäisches Patentamt**

European Patent Office

Office européen des brevets

(11) Publication number: **0 050 443**
A2

(12) **EUROPEAN PATENT APPLICATION**

(21) Application number: **81304643.0**

(22) Date of filing: **07.10.81**

(51) Int. Cl.³: **G 02 C 9/00**, G 02 B 27/00, G 01 L 1/08

---

(30) Priority: **17.10.80 US 198211**

(43) Date of publication of application: **28.04.82**
Bulletin 82/17

(84) Designated Contracting States: **AT BE CH DE FR GB IT LI LU NL SE**

(71) Applicant: **RESEARCH TRIANGLE INSTITUT, Post Office Box 12194 Research Tringle Park, Durham North Carolina (US)**
Applicant: **GALLAUDET COLLEGE, 7th Street N.E. Florida Avenue, N.E. Washington Columbia 20002 (US)**

(72) Inventor: **Cornett, Orin R., 8702 Royal Ridge Lane, Laurel Maryland 20811 (US)**
Inventor: **Beadles, Robert L., 5435 Lakeview Drive, Durham North Carolina 27712 (US)**

(74) Representative: **Harrison, David Christopher et al, MEWBURN ELLIS & CO 2/3 Cursitor Street, London EC4A 1BQ (GB)**

---

(54) **Method and apparatus for automatic cuing.**

(57) A method and apparatus for providing cues to a hearing impaired or deaf person to aid in lipreading in which phonemes are detected and analyzed to project an image into the field of view of the hearing impaired or deaf person identifying one of a plurality of groups of consonants and vowels together defining a syllable. More particularly, one of a plurality of symbols, each identifying a group of consonants is projected by a display (20) in a mode identifying a group of vowels, for example to one of four quadrants. The apparatus is preferably mounted on an eyeglass frame (24).

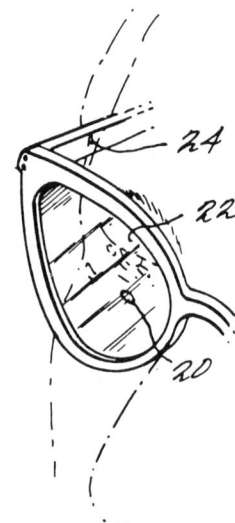

ACTORUM AG

EP 0 050 443 A2

Front page of a European patent application

The front page of a European patent application without a search report, EP 0050443 A2, is given on page 28. The features of the front page are similar to those on a British patent application but the invention is classified according to the International Patent Classification and the designated states listed are the countries in which patent protection is sought.

The amended front page of an A2 document is reprinted along with the search report as an A3 publication when the search has been carried out (see pages 30 and 31).

The European Patent Office issues ESPACE EP-A and ESPACE EP-B CD-Roms, which contain respectively European patent applications and granted paterts.

European search reports

The European Patent search report gives some detail of why each citation has been made by the patent examiner. An example is given on page 31. The kind of relevance that each citation has is given in the left hand column headed 'Category'. The codes are explained at the bottom of the page. 'X', for example, means that part at least of the application is not thought to be new, and 'Y' that it is an obvious improvement. 'A' is background information, which does not necessarily invalidate the application. The next column gives the citation to the patent or other document found by the search, and the parts of that document which are relevant to the application. The following column shows which claims in the application are affected.

The gazette

The *Bulletin* has been issued from December 1979, fortnightly until the end of 1980 and weekly from 1981 on the same day as the patent specifications. It lists separately the published applications and granted patents. The main entries, arranged by International Patent Classification, give the title of the invention, the names of the applicant and inventor and the designated states. An example is given on page 32, where the entry at G02C 9/00 for automatic cueing is for the same invention illustrated throughout this section, EP 0050443, see page 28.

There are also lists by PCT number if applicable, publication number, application number, name of applicant and designated states. The *Bulletin* provides a weekly concordance between the application number and publication number and this is cumulated on microfiche by Inpadoc (see page 32). A supplementary Cross Reference Index lists applications for European Patents made and published under the Patent Cooperation Treaty but not published as European specifications (Euro-PCT's). Information on granted European Patents designating GB is also given in the *Official Journal (Patents)* (see page 15).

Announcements of any oppositions filed against European patents and their outcome are also noted in the *Bulletin*.

(19) Europäisches Patentamt
European Patent Office
Office européen des brevets

(11) Publication number: **0 050 443**
A3

(12) **EUROPEAN PATENT APPLICATION**

(21) Application number: 81304643.0

(22) Date of filing: 07.10.81

(51) Int. Cl.³: **G 02 C 11/06**
G 02 B 27/00, G 10 L 1/08

---

(30) Priority: 17.10.80 US 198211

(43) Date of publication of application:
28.04.82 Bulletin 82/17

(88) Date of deferred publication of search report: 08.09.82

(84) Designated Contracting States:
AT BE CH DE FR GB IT LI LU NL SE

(71) Applicant: RESEARCH TRIANGLE INSTITUT
Post Office Box 12194 Research Tringle Park
Durham North Carolina(US)

(71) Applicant: GALLAUDET COLLEGE
7th Street N.E. Florida Avenue
N.E. Washington Columbia 20002(US)

(72) Inventor: Cornett, Orin R.
8702 Royal Ridge Lane
Laurel Maryland 20811(US)

(72) Inventor: Beadles, Robert L.
5435 Lakeview Drive
Durham North Carolina 27712(US)

(74) Representative: Harrison, David Christopher et al,
MEWBURN ELLIS & CO 2/3 Cursitor Street
London EC4A 1BQ(GB)

---

(54) Method and apparatus for automatic cuing.

(57) A method and apparatus for providing cues to a hearing impaired or deaf person to aid in lipreading in which phonemes are detected and analyzed to project an image into the field of view of the hearing impaired or deaf person identifying one of a plurality of groups of consonants and vowels together defining a syllable. More particularly, one of a plurality of symbols, each identifying a group of consonants is projected by a display (20) in a mode identifying a group of vowels, for example to one of four quadrants. The apparatus is preferably mounted on an eyeglass frame (24).

FIG. 1

EP 0 050 443 A3

Croydon Printing Company Ltd

Front page of a European patent application search report (subsequent to the application)

0050443

European Patent Office

EUROPEAN SEARCH REPORT

Application number

EP 81 30 4643

	DOCUMENTS CONSIDERED TO BE RELEVANT		
Category	Citation of document with indication, where appropriate, of relevant passages	Relevant to claim	CLASSIFICATION OF THE APPLICATION (Int. Cl. ³)
D,A	US-A-3 463 885 (UPTON,H.W.) *Column 2, lines 1-6; claims 1-6,11,14; figure 9* ---	1,3,5, 7,10	G 02 C 11/06 G 02 B 27/00 G 10 L 1/08
D,A	US-A-3 936 605 (UPTON,H.W.) *Abstract, lines 6-8; column 1, lines 23-30,44-45; column 2, lines 3-11,33-43,55-58; claims 19,20,21; figure 1* ---	1,3,5, 7	
A	US-A-4 117 265 (GERLACH,R.K.) *Column 1, lines 27-37; column 5, lines 3-20,65-67; claims 1,2; figure 1* -----	1,3,5, 7	

TECHNICAL FIELDS SEARCHED (Int. Cl. ³)

G 09 B
G 02 C
G 10 L
A 61 F
G 02 B

The present search report has been drawn up for all claims

Place of search THE HAGUE	Date of completion of the search 08-06-1982	Examiner SEIFERT H.U.

CATEGORY OF CITED DOCUMENTS

X : particularly relevant if taken alone
Y : particularly relevant if combined with another document of the same category
A : technological background
O : non-written disclosure
P : intermediate document

T : theory or principle underlying the invention
E : earlier patent document, but published on, or after the filing date
D : document cited in the application
L : document cited for other reasons

& : member of the same patent family, corresponding document

EPO Form 1503 03 82

European patent application search report

G02B – G03C   28.04.1982

(54) • Verbindung zwischen optischen
Fasern.
• Coupling of optical fibres.
• Epissures entre fibres optiques.
(71) SOCAPEX, 10 bis, quai Léon Blum,
F-92150 Suresnes, FR
(72) d'Auria, Luigi, F-75360 Paris Cedex
08, FR
Richin, Philippe, F-75360 Paris Cedex
08, FR
(74) Vesin, Jacques et al, THOMSON-CSF
SCPI 173, bld Haussmann, F-75360
Paris Cedex 08, FR

(51) G02B 17/08   (11) 0050584
                            A1
(86) De   (87) De
(21) 81810399.6   (22) 05.10.81
(84) AT, DE, FR, GB, IT, NL, SE
(30) 17.10.80 CH 7753/80
(54) • Spiegellinsenobjektiv von hohem
Öffnungsverhältnis.
• Wide aperture catadioptric objective
lens system.
• Lentille catadioptrique avec grande
ouverture.
(71) Canzek, Ludvik, Dr., Quellmattstrasse
3, CH-5035 Unterentfelden, CH
(72) Canzek, Ludvik, Dr., CH-5035
Unterentfelden, CH

(51) G02B 23/16   (11) 0050367
G11B 7/08

(51) G02B 27/00   (11) 0050443
G02C 9/00

(51) G02B 27/17   (11) 0050539
G01S 3/78

(51) G02C 9/00   (11) 0050443
G02B 27/00   A2
G01L 1/08
(86) En   (87) En
(21) 81304643.0   (22) 07.10.81
(84) AT, BE, CH, DE, FR, GB, IT, LI, LU,
NL, SE
(30) 17.10.80 US 198211
(54) • Verfahren und Vorrichtung zur
automatischen Anzeige.
• Method and apparatus for automatic
cuing.
• Méthode et appareil d'indication
automatique.
(71) RESEARCH TRIANGLE
INSTITUT, Post Office Box 12194
Research Tringle Park, Durham North
Carolina, US
GALLAUDET COLLEGE, 7th Street
N.E. Florida Avenue, N.E.
Washington Columbia 20002, US
(72) Cornett, Orin R., Laurel Maryland
20811, US
Beadles, Robert L., Durham North
Carolina 27712, US
(74) Harrison, David Christopher et al,
MEWBURN ELLIS & CO 2/3
Cursitor Street, London EC4A 1BQ,
GB

(51) G02F 1/133   (11) 0050357
                            A1
(86) En   (87) En
(21) 81108521.6   (22) 19.10.81
(84) DE, GB
(30) 20.10.80 JP 145809/80

(54) • Flüssigkristall-Anzeigeelement und
Verfahren zu seiner Herstellung.
• Liquid crystal display element and
process for production thereof.
• Dispositif d'affichage à cristaux
liquides et son procédé de fabrication.
(71) Hitachi, Ltd., 5-1, Marunouchi
1-chome, Chiyoda-ku Tokyo 100, JP
(72) Umeda, Takao, Hitachi-shi, JP
Igawa, Tatsuo, Sekiminamicho
Kitaibaraki-shi, JP
Simazaki, Yuzuru, Hitachi-shi, JP
Miyashita, Takao, Mito-shi, JP
Nakano, Fumio, Hitachi-shi, JP
(74) Von Füner, Alexander, Dr. K. L.
Schiff Dr. A. v. Füner et al, Dipl. Ing.
P. Strehl Dr. U. Schübel-Hopf Dipl.
Ing. D. Ebbinghaus Dr. Ing. D. Finck
Patentanwalte Mariahilfplatz 2&3,
D-8000 München 90, DE

(51) G03B 15/03   (11) 0050491
G03B 17/17   A1
(86) En   (87) En
(21) 81304832.9   (22) 16.10.81
(84) AT, BE, CH, DE, FR, GB, IT, LI, LU,
NL, SE
(30) 21.10.80 GB 8033966
(54) • Kamerazusatzbelichtungen und
Kameras.
• Improvements in or relating to
camera supplemental lighting devices
and to cameras.
• Eclairages supplémentaires de
caméras et des caméras.
(71) LEIGH INTERESTS LIMITED,
Lindon Road Brownhills, Walsall
West Midlands WS8 7BB, GB
(72) Turpin, Gerald Leslie, Great
Missenden Buckinghamshire, GB
(74) Daley, Michael John et al, F.J.
CLEVELAND & COMPANY 40/43
Chancery Lane, London, WC2A 1JQ,
GB

(51) G03B 17/17   (11) 0050491
G03B 15/03

(51) G03B 21/11   (11) 0050175
                            A1
(86) Fr   (87) Fr
(21) 80401474.4   (22) 16.10.80
(84) AT, BE, CH, DE, GB, IT, LI, LU, NL,
SE
(54) • Mikrofilmkassette und
Projektionsgerät zum Betrachten des
Mikrofilms.
• Microfilm cassette and projection
apparatus for viewing the microfilm.
• Cassette de microfilm et appareil de
projection pour la visionner.
(71) Ausseil, Dominique, 16, Allée de la
Teurtais, F-35300 Fougeres, FR
(72) Ausseil, Dominique, F-35300
Fougeres, FR
(74) Le Guen, Louis François, 13, rue
Emile Bara BP 91, F-35800 Dinard
Cedex, FR

(51) G03B 23/12   (11) 0050155*
                            A1
(86) En   (87) En
(21) 81901254.3   (22) 17.04.81
(86) US 81/00525   (86) 17.04.81
(84) AT, CH, DE, FR, GB, LI, LU, NL, SE
(87) WO 81/03075   (87) 81/25 29.10.81
(30) 18.04.80 US 141456

(54) • LESEGERÄT FÜR MIKROFICHE
IN ROLLENFORM.
• OPTICAL ROLLFICHE READER.
• LECTEUR OPTIQUE DE FICHES
EN ROULEAUX.
(71) HEADLEY, James E., 2516 Canyon
View Lane, Pasadena, CA 91107, US
(72) HEADLEY, James E., Pasadena, CA
91107, US
(74) Lewald, Dietrich, Dipl.-Ing. et al,
Birnauer Strasse 6, D-8000 München
40, DE

(51) G03B 27/62   (11) 0050508
                            A2
(86) En   (87) En
(21) 81304873.3   (22) 19.10.81
(84) DE, FR, GB
(30) 20.10.80 IT 6859780
(54) • Zuführeinrichtung für Vorlagen in
einer Kopiermaschine.
• Original feeder for copying
machines.
• Dispositif d'acheminement de
documents dans une machine à
photocopier.
(71) Ing. C. Olivetti & C., S.p.a., Via G.
Jervis 77, I-10015 Ivrea, IT
(72) De Simone, Pantaleo, I-10082
Cuorgne, IT
(74) Pears, David Ashley et al, REDDIE &
GROSE 16 Theobalds Road, London
WC1X 8PL, GB

(51) G03B 41/18   (11) 0050143
A61B 6/14

(51) G03C 1/06   (11) 0050260
                            A1
(86) De   (87) De
(21) 81107953.2   (22) 06.10.81
(84) BE, DE, FR, GB
(30) 16.10.80 DE 3039168
(54) • Fotografisches Material,
Herstellungsverfahren sowie Verfahren
zur Herstellung fotografischer Bilder.
• Photographic material, process for
its production and method for the
production of photographic pictures.
• Matériau photographique, procédé
pour sa préparation et procédé pour la
production d'images photographiques.
(71) Agfa-Gevaert Aktiengesellschaft,
Patentabteilung, D-5090 Leverkusen
1, Bayerwerk, DE
(72) von König, Anita, Dr., D-4150
Krefeld, DE
Moll, Franz, Dr., D-5090 Leverkusen
1, DE
Rosenhahn, Lothar, Dr., D-5000
Koeln 80, DE

(51) G03C 1/485   (11) 0050558
                            A2
(86) En   (87) En
(21) 81401589.7   (22) 14.10.81
(84) BE, DE, FR, GB
(30) 16.10.80 FR 8022103
(30) 17.11.80 US 207530
(54) • Photographische Emulsionen und
Elemente zur Herstellung direkt
positiver Bilder.
• Photographic emulsions and
elements capable of forming
direct-positive images.

Classified entries for published European patent applications

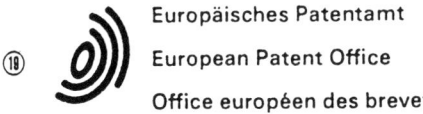

(19) Europäisches Patentamt

European Patent Office

Office européen des brevets

(11) Publication number: **0 050 443**

B1

(12) **EUROPEAN PATENT SPECIFICATION**

(45) Date of publication of patent specification: **07.05.86**

(21) Application number: **81304643.0**

(22) Date of filing: **07.10.81**

(51) Int. Cl.⁴: **G 02 C 11/06,** G 02 B 27/00,
G 10 L 5/00

(54) Method and apparatus for automatic cuing.

(30) Priority: **17.10.80 US 198211**

(43) Date of publication of application:
28.04.82 Bulletin 82/17

(45) Publication of the grant of the patent:
07.05.86 Bulletin 86/19

(84) Designated Contracting States:
BE CH DE FR GB LI NL SE

(56) References cited:
US-A-3 463 885
US-A-3 936 605
US-A-4 117 265

IEEE Transactions, Vol. ASST-23, Feb. 75, pp. 54-67

IEEE Transactions, Vol. ASST-24, Apr. 76, pp. 170-182

(73) Proprietor: **RESEARCH TRIANGLE INSTITUTE**
Post Office Box 12194 Research Tringle Park
Durham North Carolina (US)

(73) Proprietor: **GALLAUDET COLLEGE**
7th Street N.E. Florida Avenue
N.E. Washington Columbia 20002 (US)

(72) Inventor: **Cornett, Orin R.**
8702 Royal Ridge Lane
Laurel Maryland 20811 (US)
Inventor: **Beadles, Robert L.**
5435 Lakeview Drive
Durham North Carolina 27712 (US)

(74) Representative: **Harrison, David Christopher**
et al
MEWBURN ELLIS & CO 2/3 Cursitor Street
London EC4A 1BQ (GB)

Note: Within nine months from the publication of the mention of the grant of the European patent, any person may give notice to the European Patent Office of opposition to the European patent granted. Notice of opposition shall be filed in a written reasoned statement. It shall not be deemed to have been filed until the opposition fee has been paid. (Art. 99(1) European patent convention).

EP 0 050 443 B1

Courier Press, Leamington Spa, England.

Front page of a European granted patent

The name indexes in the weekly *Bulletin* are cumulated into two annual indexes: to published applications, and to granted patents. The indexes include inventors as well as applicants.

Abstracts

European patent abstracts (formerly *European patents report*) is published by Derwent Publications Ltd. It provides English language abstracts and bibliographic information for all published applications for European patents ('A' specifications). First issues appeared with abstracts in specification number order but from early 1981 abstracts were arranged instead in Derwent's own classified order. In late 1981, three separate classified issues – Chemical, Electrical, and General/Mechanical – were produced and these were joined by an additional series for 'A' specifications in numerical order beginning in 1985. A series for granted patents was published from 1985 - 1990.

The official abstracts in English, French or German taken from the front pages of the 'A' specifications are republished in classified sequence by Wila Verlag as *Auszüge aus den Europäischen Patentanmeldungen*.

Translations

Some designated states require applicants to submit translations into the official language of the country of claims in European Patent applications and of the entire granted European patent before protection becomes fully effective in that state. The Patents (Amendment) Rules 1987 introduced this provision in the United Kingdom for applications published and patents granted after 1 September 1987.

Case law

Appeals can be lodged against decisions made by the European Patent Office and will be examined by Appeal Boards. SRIS receives copies of the texts of all decisions of the Boards of Appeal in the language of the hearing (English, French or German). Certain of these decisions are subsequently published in the Official Journal of the EPO.

SRIS and the European Patent Convention patents

This section explains about procedures followed by SRIS in making European Patent Convention patent information available.

SRIS displays two sets of the current week's published patent applications on a table. They are in numerical order, but since this is not a classified order, subject, as well as name access is only possible by using the *Bulletin*. The last three weeks' 'B' specifications are kept in separate boxes at SRIS to facilitate searching.

Access to the official European Patent Office registers, giving stages of progress of applications, can be made direct to the European Patent Office or by means of an online search. The registers give information about a patent application up to and including grant and through the opposition procedure, if invoked.

SRIS maintains non-official records taken from the *European Patent Bulletin* in the form of the EPO publications register. An example is given on page 36. The entries in this register show, against the publication number, the date of shelving in SRIS (generally the date of the publication); the date of shelving of any search reports published separately and amendments; the date of shelving of the granted patent; and the date of shelving of any translations. The date of an announcement in the *Bulletins* of any withdrawal or refusal and the numbers of any prior PCT or later GB publication are added. To find the exact dates it is necessary to check the *Bulletin* issue for that date.

Registers showing the status of granted European patents designating GB can be inspected at the UK Patent Office or on OPTICS, its online database.

The Inpadoc service, both in the form of microfiche and online services, give status information on European Patents (see pages 77, 86).

A card index to all decisions of the EPO Boards of Appeal is available for reference at SRIS.

Further reading

How to get a European Patent. European Patent Office

National law relating to the EPC. European Patent Office.

Updating the European Patent Convention. E. Armitage. *IIC*, 1991. 22 (1), 1-10.

The dynamism of the European patent system. P. Braendli. *IIC*, 1991. 22 (2), 177-194.

THIS INDEX TO EPO SPECIFICATIONS IS COMPILED BY THE SCIENCE REFERENCE LIBRARY AND IS NOT AN OFFICIAL RECORD (The Register of European Patents is also accessible on-line).

E.P.O. PUBLISHED APPLICATION NUMBER	Date Shelved	AMENDMENTS REPRINTS ETC.	SEARCH REPORT PUBLISHED SEPARATELY Date Shelved	APPLICATION WITHDRAWN See E.P. Bulletin dated	GRANTED SPECIFICATION Date Shelved	P.C.T. PUBLICATION NUMBER	REMARKS
0197911	(illegible)			6 APR 1988			TRANSLATION OF ... 23/5/90 B SPEC SHELVED
2			29 MAY 1987		11 APR 1990		
3					25 JUL 1989		
4							TRANSLATION OF 22/6/88 B SPEC SHELVED
5			10 AUG 1988		25 MAY 1988		B SPEC SHELVED 7/3/90
6					14 FEB 1990		B SPEC SHELVED 16.11.88
7			24 AUG 1988		28 SEP 1988		
8	(illegible)			8 NOV 1989			
9	N/P				26 NOV 1987		TRANSLATION OF 23/12/87 B SPEC SHELVED
0197920				12 JUL 1989	31 OCT 1990	WO85/02831	
1				1 APR 1987		WO86/02260	
2				6 MAY 1987		WO86/02161	
3				27 FEB 1991		WO86/02328	
4				15 APR 1987		WO86/02292	
5				13 MAY 1987		WO86/02294	
6				8 APR 1987		WO86/02300	
7				8 APR 1987		WO86/02514	
8			10 AUG 1988	29 APR 1987		WO86/02758	
9				9 SEP 1987		WO86/02626	
0197930				23 DEC 1986	27 JAN 1988	WO86/02522	
1				25 MAR 1987		WO86/02751	
2				25 MAR 1987		WO86/02775	
3						WO85/02389	
4				24 JUN 1987	22 AUG 1990	WO85/02807	
5				8 APR 1987		WO85/02452	
6						WO85/03160	
7				29 APR 1987		WO85/02866	
8				26 ... 1991		WO86/02277	
						WO86/02663	

Non-official European Publication Register

4 PATENT COOPERATION TREATY PUBLICATIONS

The Patent Cooperation Treaty (PCT) provides, on the basis of a single 'international' application in one language, for an international search which will be effective in any of the countries which are party to the treaty.

The first PCT applications were filed on 1 June 1978 and publication began in October 1978. The number of PCT documents is rapidly increasing. In 1990 over 16,000 applications were published. The PCT is administered by the World Intellectual Property Organization (WIPO) in Geneva.

Currently the member countries are:

Australia	(AU)	Korea, Dem Rep	(KP)
Austria	(AT)	Korea, Rep of	(KR)
Barbados	(BB)	Liechtenstein	(LI)
Belgium	(BE)	Luxembourg	(LU)
Benin	(BJ)	Madagascar	(MG)
Brazil	(BR)	Malawi	(MW)
Bulgaria	(BG)	Mali	(ML)
Burkina Faso	(BF)	Mauritania	(MR)
Cameroon	(CM)	Monaco	(MC)
Canada	(CA)	Mongolia	(MN)
Cent African Rep	(CF)	Netherlands	(NL)
Chad	(TD)	Norway	(NO)
Congo	(CG)	Poland	(PL)
Czechoslovakia	(CS)	Romania	(RO)
Denmark	(DK)	Senegal	(SN)
Finland	(FI)	Soviet Union	(SU)
France	(FR)	Spain	(ES)
Gabon	(GA)	Sri Lanka	(LK)
Germany	(DE)	Sudan	(SD)
Greece	(GR)	Sweden	(SE)
Guinea	(GN)	Switzerland	(CH)
Hungary	(HU)	Togo	(TG)
Italy	(IT)	United Kingdom	(GB)
Ivory Coast	(CI)	United States	(US)
Japan	(JP)		

Numeration

Patent applications are given a number such as PCT/US86/02021 when they are filed at a receiving office on behalf of the Patent Cooperation Treaty. US indicates that it was filed via the United States Patent and Trademark Office, 86 that it was filed in 1986, and 02021 is the consecutive number that began from 1 for 1986 within the 'US' sequence.

The published application is numbered e.g. WO 87/06048. WO is the country code for the PCT, and 87 indicates that it was published in 1987. 06048 is the consecutive number that began from 1 for this application during the year.

Patenting procedure

Applications for patents are published every fortnight as PCT 'A' specifications in either English, French, German, Japanese, Spanish or Russian with an English language abstract and generally a search report 18 months after the priority date.

After publication and assessment of the search report the applicant may abandon the application or proceed to obtain, as appropriate, a European patent and/or selected national patents in the states which were designated.

Although there is a procedure for obtaining an international preliminary examination under the PCT it is ultimately always necessary to apply separately to the national or regional patent offices for the grant of a patent, see page 39.

Specifications

As with the European patent specification, the front page of the PCT specification contains bibliographic information, a list of designated states, an abstract and a significant drawing. The search report appears at the back of the specification and is similar to the European Patent Convention search reports.

The specifications are arranged by International Patent Classification before numbering. The kind of document code which is given on the front page of the specification has the following significance:

Al   Publication of international application with international search report.

A2   Publication of international application without international search report.

A3   Later publication of international search report together with a revised specification front page.

A front page of a PCT application is shown on page 40.

PCT Applications are now published on CD-ROM as well as on paper in the WORLD series issued by the European Patent Office.

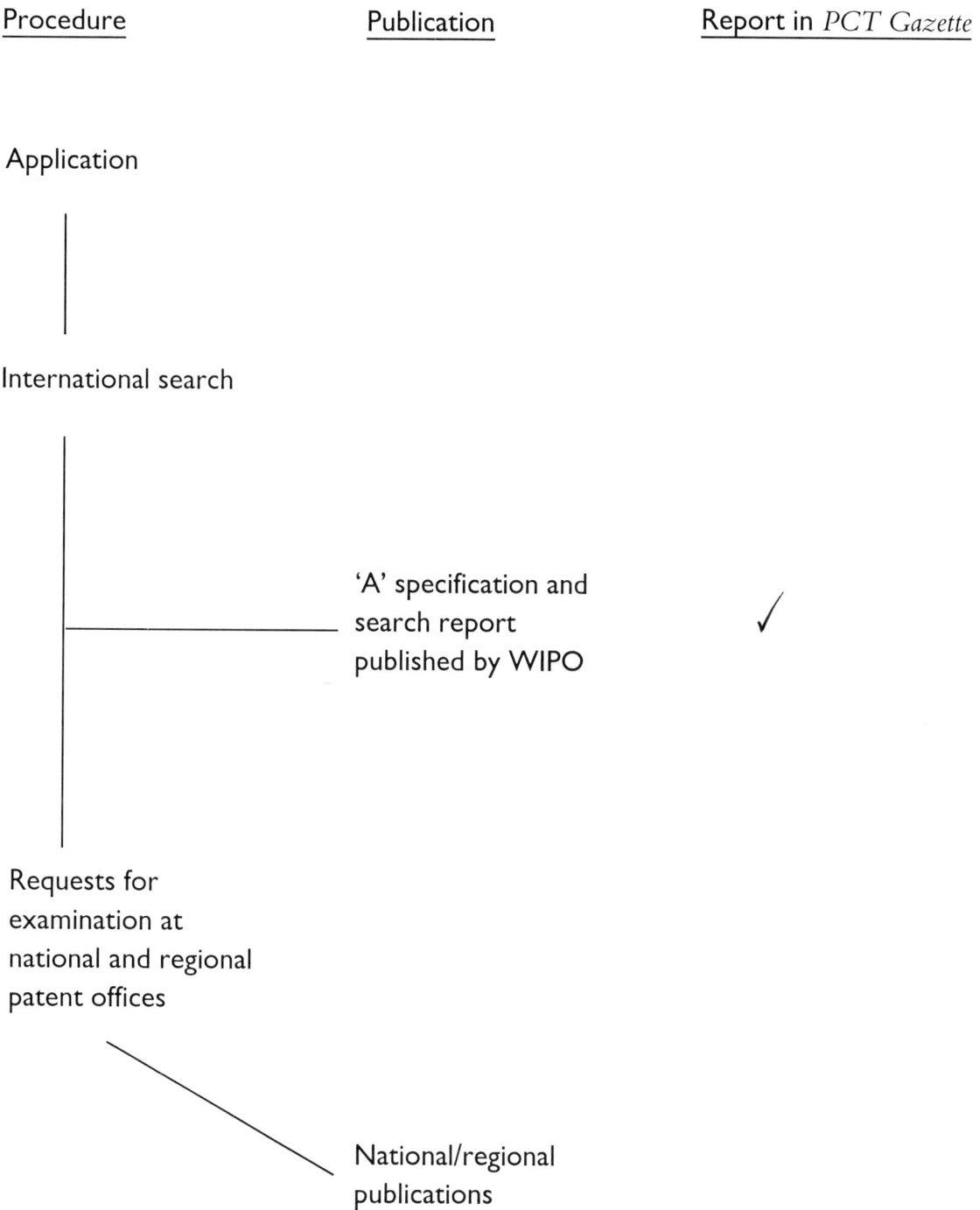

Procedure	Publication	Report in *PCT Gazette*
Application		
International search		
	'A' specification and search report published by WIPO	✓
Requests for examination at national and regional patent offices		
	National/regional publications	

Progress of an application under the PCT

PCT   WORLD INTELLECTUAL PROPERTY ORGANIZATION
International Bureau

INTERNATIONAL APPLICATION PUBLISHED UNDER THE PATENT COOPERATION TREATY (PCT)

(51) International Patent Classification ³ : H01K 1/62	A1	(11) International Publication Number: **WO 84/ 00080**
		(43) International Publication Date: 5 January 1984 (05.01.84)

(21) International Application Number: PCT/US83/00916

(22) International Filing Date: 13 June 1983 (13.06.83)

(31) Priority Application Number: 387,737

(32) Priority Date: 14 June 1982 (14.06.82)

(33) Priority Country: US

(71) Applicant: DIOLIGHT TECHNOLOGY, INCORPO-RATED [US/US]; 70 East Long Lake Road, Bloomfield Hills, MI 48013 (US).

(72) Inventors: KEATING, Kevin, J. ; 1098 Northlawn, Birmingham, MI 48010 (US). MONAHAN, Russell ; 1143 Sweet Road, Ypsilanti, MI 48197 (US).

(74) Agent: ETHINGTON, Paul, J.; Reising, Ethington, Barnard, Perry & Milton, 3290 West Big Beaver Road, Troy, MI 48084 (US).

(81) Designated States: AU, BR, DE, DE (European patent), FR (European patent), GB, GB (European patent), JP, KP.

Published
With international search report.
With amended claims.

(54) Title: EXTENDED LIFE INCANDESCENT LAMP WITH SELF-CONTAINED DIODE AND REFLECTOR

(57) Abstract

An incandescent lamp having an extended operating life with a high operating efficiency in terms of lumens per watt. The lamp incorporates a rectifying diode (40) in series with the filament (26) and the filament (26) has an operating temperature less than 2550°C. A reflector (28) is mounted in the neck portion (14) of the lamp for reflecting visible light outwardly through the bulb portion (12) of the envelope (10) and for minimizing convection currents in the base (32) of the lamp. The diode (40) is disposed between the reflector (28) and the lamp base (32).

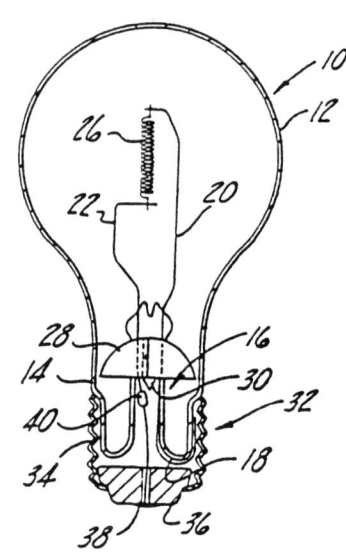

Front page of a PCT patent application

PCT Gazette

The fortnightly *PCT Gazette* contains details of PCT specifications which are published on the same day as the *Gazette*. The entries are arranged in order of publication number and comprise an abstract and a drawing (where applicable) together with full bibliographic information (see example on page 42).

There are also applicant's (but not inventor's) name indexes, class indexes, concordance tables relating application numbers to publication numbers and lists of publication numbers by designated country. There are also annual name, class and concordance indexes.

Abstracts

The front page information (including the English abstract) is published every fortnight in reduced format in the *PCT Gazette,* see above.

In addition, Derwent Publications publishes *PCT Patent Abstracts* every fortnight. It provides English language abstracts and bibliographic information for all published PCT applications. The abstracts are published in Derwent's own classified order but with a numerical concordance.

SRIS and the Patent Cooperation Treaty

This section explains about procedures followed by SRIS in making British patent information available.

SRIS displays the current fortnight's patent applications on a table to help current awareness searchers. They are in classified order so it is easy to find material on a particular topic.

SRIS maintains a non-official register giving the stages of progress of PCT applications. The information in this register is taken from the *PCT Gazette* and recorded in the form of a PCT publications register (bound in blue). Against the publication number are the date of shelving in SRIS and that of any subsequent amendments as well as any eventual EP or GB publication numbers. The separate registers for those patent systems should then be consulted to find what happened to the application. In this way the progress of any attempt to secure protection for the invention in Britain can be determined.

Further reading

PCT Applicants' Guide. World Intellectual Property Organization.

PCT questions in the Euro-PCT context. G. Gall. *European Intellectual Property Review,* Nov 1984. 6 (11), 302-310.

The Patent Cooperation Treaty: ten years of implementation. B. Bartels, D. Bouchez, P. Higham. *World Patent Information,* 1988. 10 (2), 99-103.

H01K

(21) Int. Application Number: PCT/US83/00916	(51) International Patent Classification [3] :		(11) Int. Publication Number: **WO 84/00080**
(22) Int. Filing Date: 13 June 1983 (13.06.83)	**H01K 1/62**	**A1**	(43) Int. Publication Date: 5 January 1984 (05.01.84)

(31) Priority Application Number: 387,737

(32) Priority Date: 14 June 1982 (14.06.82)

(33) Priority Country: US

(71) Applicant: DIOLIGHT TECHNOLOGY, IN-CORPORATED [US/US]; 70 East Long Lake Road, Bloomfield Hills, MI 48013 (US).

(72) Inventors: KEATING, Kevin, J. ; 1098 Northlawn, Birmingham, MI 48010 (US). MONAHAN, Russell ; 1143 Sweet Road, Ypsilanti, MI 48197 (US).

(74) Agent: ETHINGTON, Paul, J.; Reising, Ethington, Barnard, Perry & Milton, 3290 West Big Beaver Road, Troy, MI 48084 (US).

(81) Designated States: AU, BR, DE, DE (European patent), FR (European patent), GB, GB (European patent), JP, KP.

Published
With international search report.
With amended claims.

(54) Title: EXTENDED LIFE INCANDESCENT LAMP WITH SELF-CONTAINED DIODE AND REFLECTOR

(57) Abstract

An incandescent lamp having an extended operating life with a high operating efficiency in terms of lumens per watt. The lamp incorporates a rectifying diode (40) in series with the filament (26) and the filament (26) has an operating temperature less than 2550°C. A reflector (28) is mounted in the neck portion (14) of the lamp for reflecting visible light outwardly through the bulb portion (12) of the envelope (10) and for minimizing convection currents in the base (32) of the lamp. The diode (40) is disposed between the reflector (28) and the lamp base (32).

H01L

(21) Int. Application Number: PCT/US82/00822	(51) International Patent Classification [3] :		(11) Int. Publication Number: **WO 84/00081**
(22) Int. Filing Date: 14 June 1982 (14.06.82)	**H01L 41/22; B23K 26/06**	**A1**	(43) Int. Publication Date: 5 January 1984 (05.01.84)

(71) Applicant: GTE PRODUCTS CORPORATION [US/US]; One Stamford Forum, Stamford, CT 06904 (US).

(72) Inventors: CLAES, Roger ; B-3800 St. Truiden (BE). VANNOPPEN, Jean ; B-3300 Tienen (BE).

(74) Agent: ODOZYNSKI, John, A.; GTE Service Corporation, Precision Materials Group, 100 Endicott Street, Danvers, MA 01923 (US).

(81) Designated States: AT (European patent), BE (European patent), CH (European patent), DE (European patent), FR (European patent), GB (European patent), JP, LU (European patent), NL (European patent), SE (European patent).

Published
With international search report.

(54) Title: APPARATUS FOR TRIMMING OF PIEZOELECTRIC COMPONENTS

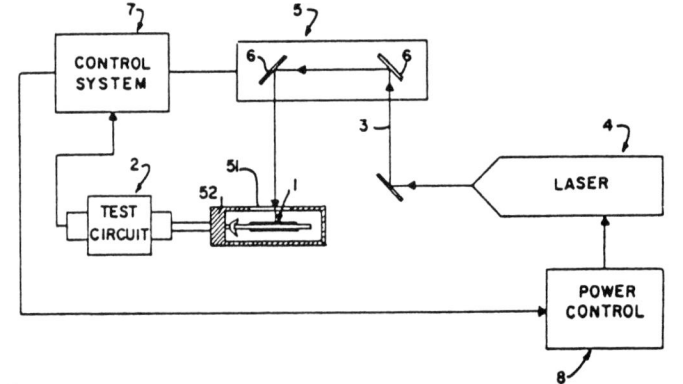

(57) Abstract

An apparatus for trimming piezoelectric components of the type characterized by a piezoelectric substrate (41) upon which is deposited a conductive material (42). The apparatus includes a housing (52) for the device at least a portion of which (51) is substantially transparent to electromagnetic energy at a predetermined wavelength. A laser (4) provides energy at the predetermined wavelength and with sufficient intensity to evaporate the conductive material. A test circuit coupled (2) to the device monitors specific frequency-related characteristics and provides a corresponding output to a control system (7). The control system (7) dictates the operation of a deflection system that directs the electromagnetic energy provided by the laser means (4) and a laser power control device (8) that regulates the intensity of that energy.

Page from the PCT Gazette

5 UNITED STATES PATENT PUBLICATIONS

Numeration

American patent applications are given application (or, in the gazette, 'ser.') numbers in a consecutive sequence from 1 to under 1,000,000, with the number reverting to 1 in the year when the numbers seem likely to go over 1,000,000, in other words every nine years or so. The numeration last reverted to 1 in 1987.

On publication (at grant) the patents are numbered consecutively in a series from 1 in 1836 to the present day. The first patent issued in 1991 was numbered 4,980,927.

Patenting procedure

The United States is now alone (with the Philippines) in establishing novelty by the date of invention rather than that of filing. This leads to carefully documented notebooks in American research establishments which show when an invention was first thought of. American patents can also be modified while pending by adding further novel information. These are reflected by 'continuation in part' wordings in code 62 on the front pages.

The inventor, called the 'patentee', is considered to be the owner of the intellectual property in the invention, and the applicant, the company who will manufacture it, is called the 'assignee'. This occurs even if the inventor works for the assignee.

Patents are only published if granted, rather than at an application stage as well.

Until recently there used to be a delay of several years before the patent was published, but the average delay until 'disposal' (rejecting or granting a patent) is now said to be 18 months. The systems that published applications before grant, at 18 months, had previously often had an advantage in providing early information. This speed-up in American examining has meant that this is no longer so evident.

About 100,000 patents are currently published annually.

Specifications

The prior art, description and drawings in an American patent are often more detailed than in similar patents abroad. This is due to American legal requirements.

The front page of an American patent is shown on the opposite page. It is clearly set out, as is the text of the patent.

The patent was published as 4,718,426. The date of grant is given as 12 January 1988. The inventor's name, Russell, is given at the top left.

Only the town of the inventor (or patentee) and of the applicant (or assignee) must be given at codes 73 or 75, although sometimes full addresses are provided.

The application number for this patent is given in code 21 as 6,626. 'Related US application data' is helpful in linking up directly related material by citing filing or published numbers. Because of the nature of American patent laws many applications are split up in this way.

The national classification numbers for this patent are given at code 52. This is taken from the highly detailed scheme which emphasizes applications, or products, rather than ideas. The IPC numbers on the patents are derived from a computer concordance to the American classification, and are often regarded as being somewhat unreliable. They frequently refer to a single, broader topic rather than to two or more more detailed concepts.

The 'references cited' at code 56 constitute the search report. This information has been given in American patents and designs since 1947. Against each patent is given the US classification number which has caused the examiner to cite the patent on the grounds of relevance. It is common for examiners to cite scientific papers as well as patents, as has happened in this case.

The patents are protected for 17 years from the date of grant, and until recently could never lapse, since no renewal fees were paid. Patents which were applied for after 12 December 1980, however, must have 'maintenance fees' paid at intervals. There is no compulsion to make the applicant 'work' the invention.

MicroPatent's Patent Images and Research Publication's Patent View are two CD-Rom products containing the complete text of American patents.

The gazette

The *Official Gazette* is published weekly. The patents in each issue are divided into three sections: General and Mechanical; Chemical; and Electrical. Within each section they are arranged in US classification order. Since the classification does not neatly divide into the three subject areas it means that there are effectively three classified sequences in each issue.

The patents are numbered after they have been classified so that they form a single numerical sequence. An example from the *Gazette* is given on page 45.

United States Patent [19]

Russell

[11] Patent Number: **4,718,426**

[45] Date of Patent: **Jan. 12, 1988**

[54] **METHOD FOR DETERMINING DIASTOLIC ARTERIAL BLOOD PRESSURE IN A SUBJECT**

[75] Inventor: **Ted W. Russell,** Northport, N.Y.

[73] Assignee: **Cortronic Corporation,** Ronkonkoma, N.Y.

[21] Appl. No.: **6,626**

[22] Filed: **Jan. 22, 1987**

Related U.S. Application Data

[62] Division of Ser. No. 581,134, Feb. 17, 1984, Pat. No. 4,669,485.

[51] Int. Cl.⁴ A61B 5/02
[52] U.S. Cl. 128/679; 128/681; 128/677
[58] Field of Search 128/672, 677–686

[56] **References Cited**

U.S. PATENT DOCUMENTS

3,157,177	11/1964	Smith	128/679
3,348,534	10/1967	Marx et al.	128/679
3,581,734	6/1971	Croslin	128/679
3,903,872	9/1975	Link	128/681
4,009,709	3/1977	Link et al.	128/681
4,074,711	2/1978	Link et al.	128/681
4,105,021	8/1978	Williams et al.	128/683
4,216,779	8/1980	Squires et al.	128/682
4,245,648	1/1981	Trimmer et al.	128/672 X
4,262,675	4/1981	Kubo et al.	128/680
4,271,843	6/1981	Flynn	128/681
4,271,844	6/1981	Croslin	128/681
4,313,445	2/1982	Georgi	128/680
4,343,314	8/1982	Sramek	128/680
4,349,034	9/1982	Ramsey, III	128/681
4,378,807	4/1983	Peterson et al.	128/677
4,407,297	10/1983	Croslin	128/681
4,408,614	10/1983	Weaver et al.	128/680
4,418,700	12/1983	Warner	128/672 X
4,437,469	3/1984	Djordjevich et al.	128/677 X
4,479,494	10/1984	McEwen	128/682 X

FOREIGN PATENT DOCUMENTS

0517480	1/1972	Switzerland .	
2092309	8/1982	United Kingdom	128/672

OTHER PUBLICATIONS

D. Bahr et al.; "The Automatic Arterial Tonometer"; *Engr. in Med. and Biol.-Proc. of the Annual Conference,* 1973, vol. 15, p. 259.
D. Bergel; "The Dynamic Elastic Properties of the Arterial Wall"; *Journal of Physiology,* (1961), vol. 156, pp. 458–469.
D. Bergel; "The Static Elastic Properties of the Arterial Wall"; *Journal of Physiology,* (1961), vol. 156, pp. 445–457.

(List continued on next page.)

Primary Examiner—Kyle L. Howell
Assistant Examiner—Angela D. Sykes
Attorney, Agent, or Firm—Stanger, Michaelson, Reynolds, Spivak & Tobia

[57] **ABSTRACT**

Apparatus and related methods for continuous long-term non-invasive measurement of the pressure of a pulsatile fluid flowing through a flexible tube, particularly human arterial blood flow, is disclosed. Specifically, the apparatus provides a continuous calibrated pressure measurement by first undertaking a "calibration" phase comprised of determining the pressure at various pre-defined conditions of flow and, in response thereto, ascertaining the values of a plurality of coefficients each of which is associated with a corresponding term in a pre-defined function that characterizes fluid pressure in relation to pulsatile displacement of the wall of the tube; and second, undertaking a "continuous monitoring" phase comprised of determining each subsequently occurring pressure value as the pre-defined function of each corresponding pulsatile wall displacement value, and re-initiating the calibration phase at the expiration of pre-defined time intervals which adaptively change based upon current and prior results. Methods, which are particularly useful in conjunction with the disclosed apparatus, for ascertaining systolic and diastolic arterial blood pressure values are also disclosed.

5 Claims, 59 Drawing Figures

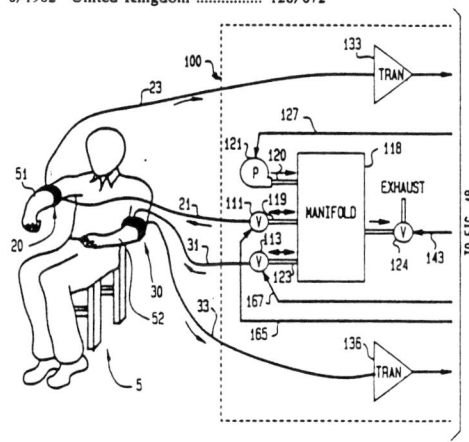

Front page of an American granted patent

4,729,445

CATERPILLAR CHASSIS FOR HEAVY VEHICLES

Horst Kolleth, Zeltweg, Austria, assignor to Voest-Alpine Aktiengesellschaft, Austria
Filed Oct. 27, 1986, Ser. No. 923,644
Claims priority, application Austria, Oct. 28, 1985, 3114/85
Int. Cl.⁴ B62D 55/00, 11/04
U.S. Cl. 180—9.46   4 Claims

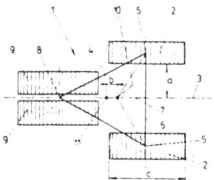

1. In an endless tread chassis for a heavy vehicle of the kind which includes a rotary table swingable about a vertical swing axis for carrying a cantilever arm, said chassis having a longitudinal axis lying in a longitudinal vertical plane and having at least four endless treads, the first and second of said treads being parallel and located on opposite sides of said longitudinal axis in a first plane which is transverse to said longitudinal axis and the third and fourth of said treads being parallel and located on opposite sides of said longitudinal axis at greater distances from said longitudinal axis than the first and second treads and in a second plane which is transverse to said longitudinal axis and which is spaced along said longitudinal axis from said first plane, said first and second treads being connected to the chassis at a common first connecting point lying in said longitudinal vertical plane and said third and fourth treads being connected to the chassis at second and third connecting points laterally spaced from said longitudinal vertical plane, said first, second and third connecting points defining the apexes of a supporting triangle and said vertical swing axis being located away from the center of gravity of said supporting triangle in a direction toward said common first connecting point.

4,729,446

MOBILE SPHERE

John S. Sefton, 280 Angust Crescent, Regina, Saskatchewan, Canada S4T 6N4
Filed Oct. 27, 1986, Ser. No. 923,406
Claims priority, application Canada, Oct. 31, 1985, 494298
Int. Cl.⁴ A63G 29/00
U.S. Cl. 180—21   20 Claims

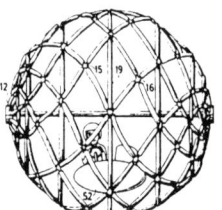

5. A hollow sphere for use as a toy, a recreational device or as a transportation vehicle comprising in combination one set of resilient members extending from one end of an axis to the other and spiralling through 360°, another set of resilient members extending from said other end of said axis to said one end thereof and spiralling through 360° in the opposite direction of said one set, each set including a plurality of resilient members secured by one end thereof to a hub at said one end of said axis

and by the other ends thereof to a hub at said other end of said axis, the members of said one set being secured to the members of the other set where said members intersect thus forming a geodesic configuration, an axle extending between said hubs, a frame bearably supported upon said axle and depending therefrom, a source of power carried by said frame, including a foot pedal assembly depending below said frame means connecting said source of power to said axle for rotating same and hence rotating said sphere, and passenger supporting means attached to said frame operatively adjacent said foot pedal assembly.

4,729,447

SAFEGUARD FOR POWERED WHEELCHAIR

John F. Morse, 1699 Hines Hill Rd., Hudson, Ohio 44236
Filed Sep. 15, 1986, Ser. No. 907,851
Int. Cl.⁴ B62D 61/08
U.S. Cl. 180—65.1   6 Claims

1. In combination with an electric powered wheelchair having a steering column forward of the chair seat, a handle bar mounted on the top of said column and extending substantially crosswise of the longitudinal axis of the wheelchair, a switch box mounted at the top of the steering column, and a switch control lever pivoted on the underside of said box spaced behind said handle bar and extending substantially parallel therewith, the improvement comprising a fail-safe guard mounted on said handle bar and extending rearwardly beneath said control lever to prevent accidental contact therewith by the body of the chair occupant when rising to dismount.

4,729,448

OFFSET TRUNNION BRACKET ON STEER DRIVE AXLE

James L. Sullivan, Rochester, Mich., assignor to Rockwell International Corporation, Pittsburgh, Pa.
Filed Jan. 30, 1987, Ser. No. 8,996
Int. Cl.⁴ B62D 5/06
U.S. Cl. 180—161   6 Claims

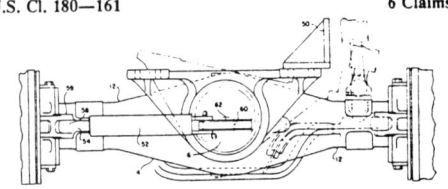

1. Steer-driver axle housing apparatus for supporting a trunnion bracket comprising a differential-receiving bowl centered around a first vertical plane extending fore and aft through the bowl and a first horizontal plane extending through the bowl, first and second axle housing arms extending laterally oppositely from the differential-receiving bowl, first and second steering knuckles respectively mounted on opposite ends of the axle housing arms remote from the bowl, steering arms con-

Page from the US Official Gazette

The *Gazette* gives brief bibliographic details for each patent, together with a representative illustration and usually a single claim, which in effect forms an abstract. Usually the first claim is given, but sometimes a later claim is considered to be more useful in indicating the inventive step.

Each issue has a single index of the patentees and assignees, and another arranged by US classification.

These indexes have references from the assignees to the patentees. The entry for the patentees give the name of any assignee. Annual indexes are also published.

Abstracts

The United States does not publish abstracts of its patents. Bibliographic details and claims are published in numerical order in its gazette.

Derwent Publications publishes *United States Patents Abstracts* weekly in three sections: Chemical; Mechanical and general; and Electrical. They provide bibliorgaphic details, an abstract and if relevant a drawing for each patent. The abstracts are published in Derwent's own classified order but with a numerical concordance.

Other forms of patent documents

American patents are often called utility patents to distinguish them from design patents, which are for the appearance of an object.

Plant patents (coded PP) are also published, the United States being the only country to allow patents for new plant varieties.

If a published American patent is found to be inaccurate, due to, for instance, printing errors, then the patent and its number is replaced by a Reissue (Re) patent. A reference will be found at the original document. Alternatively a correction slip is attached to the original.

If, on the other hand, through opposition in the courts the patent is amended then a reexamination certificate is published, which in SRIS is attached to the original. These are designated as B1 or B2 documents.

There are disclosures, where the applicant describes the invention in such detail that neither he nor anyone else can take out a patent. They used to be called Defensives (coded T), and from 1984 are called Statutory Inventions. Their designation is H.

Claims and bibliographic details for all these categories are published in the gazette.

SRIS and the United States patents

A microfiche concordance is available at SRIS for filings from 1968, although it is not up to date, and more recent publications can be traced on CD-ROM, using the APS discs.

Patent applications from US government agencies (but lacking the claims and the front page information) are received at SRIS and can often reduce further the delay in seeing the information. They are entitled 'US government owned applications' at SRIS and are numbered by the application number preceded by a number indicating the sequence of filing numbers running from 1 to one million. A recent number in this series, for example, is 7 - 618201.

Further reading

Short-run trends in United States patent activity. G. Brunk, G. Demack. *Scientometrics*, July 1987, 12 (1/2), 111-133.

Reissue and reexamination, C. Hamre. *Idea*, 1989. 29 (3/4), 311-331.

Patent law harmonization treaty decision is not far off: what course should the US take? W. Fryer III. *Idea*, 1990. 30 (4), 309-354.

Entering our third century, H. Manbeck Jr. *Patent Office Society Journal*, Dec 1990. 72 (12), 1177-1187 [Discusses possible changes in the US Patent and Trademark Office in the future].

Statement of Donald W. Banner on Patent and Trademark Office oversight and reauthorization, D. Banner. *Patent Office Society Journal*, July 1991. 73 (7), 543-556 [Testimony before a House subcommittee by a former Commissioner of the Office].

6 GERMAN PATENT PUBLICATIONS

Numeration

The numbers given to German patent applications do not change when they are published. Therefore no concordances are needed. From 1968 German patents were given numbers beginning with 1800001. The first two digits advance by one each year. By adding 50 to the first two digits it is possible to find the year of filing, so that numbers for patents filed in 1990 begin with the digits 40.

Two extra batches of numbers are used by the German Patent Office to designate European Patent Convention and Patent Cooperation Treaty applications. Within each year, numbers within the 60,001 onwards range indicate European Patent Convention applications that designate Germany, in translation if necessary. The second range consists of numbers beginning with 90,001. These designate Patent Cooperation Treaty applications that designate Germany, in translation if necessary.

Patenting procedure

There have been some changes in the stages of publication. Until 1957 only the granted patent, or Patentschrift, was ever published. German documentation often refers to the stages with acronyms, and these are called PS.

From 1957 applications were published before the granted specification as an Auslegeschrift (AS). From 1968 a third stage, preceding the Auslegeschrift, was introduced. This is the Offenlegungsschrift (OS). These are still being published, but the AS were discontinued after 1981 legislation.

Therefore only two stages are published today. Speedily processed applications, however may be published only once as a Patentschrift, 18 months after the priority date.

The applicant has up to seven years from the publication of the OS document to decide whether or not to ask for an examination. Because of the potential delay, there is much interest in whether or not a German application has been withdrawn or is still pending. German patents are protected for a maximum of 20 years from the date of filing. About 20,000 patents are granted annually.

Germany grants utility models as well as patents. These are called Gebrauchsmuster (GM). These are simpler inventions in the non-chemical fields. The complete texts are not held at SRIS. They are not examined, and are published about six weeks after filing. They are numbered within each year, e.g. for 1987 8706544 and so on. Protection is given for a maximum of 10 years.

Specifications

The front page of a typical Offenlegungsschrift is shown on page 51. It was filed on 13 August 1987 (code 22, Anmeldetag) when it was given the number 3727052. The Al signifies that it is the first time that the patent has ever been published, with A signifying an application, and the 1 showing that it is the first time the application has been published. The INID codes can be used if necessary to identify the other data. Code 30, for instance, gives the Japanese priority data. No search report is published at this stage. The claims, which are at the end of the description (not shown) are called the Patentanspruche.

Page 52 shows the front page of a Patentschrift. It is designated as a C2. The C signifies that it is theoretically the third stage of publication, despite the disappearance of the Auslegeschrift. The 2 shows that it is being published for the second time. If the Patentschrift was the only publication then it would be a Cl. The search report is given against code 56. It consists of a scientific paper, three US patents, one German patent and a 'GM', or a Gebrauchsmuster (a German utility model).

German OS, PS, GM and European Patent Convention translations are available on ESPACE-DE, a CD-ROM product from Bertelsmann.

Abstracts

There are a number of gazettes, all of which are published weekly. They give claims rather than abstracts.

The *Patentablatt* lists in IPC order bibliographic details of published patents. Cumulated indexes of applicants are published twice a year. The *Auszüge aus den Offenlegungschriften* is divided into three series, Teil 1 to 3, according to the subject matter. They consist of classified arrangements of the main claims, with a drawing, of the Offenlegungschrift documents. An example is on page 53.

The *Auszüge aus den Patentschriften* is a similar publication. Since many documents are published for the first time as Patentschriften it is a useful series for alerting purposes.

The utility models have a similar publication, the *Auszüge aus den Gebrauchsmustern*.

Derwent Publications publishes the *German Patents Abstracts* (previously *German Patents Gazette*), in three subject series, which cover the Offenlegungschriften, and the *German Patents Abstracts*, examined series, in two series, for the Patentschriften.

(19) **BUNDESREPUBLIK DEUTSCHLAND**

DEUTSCHES PATENTAMT

(12) **Offenlegungsschrift**

(11) **DE 37 27 052 A1**

(51) Int. Cl. 4:
G 03 D 13/00
// G03B 27/30,
H04N 1/23,B41J 3/21

DE 37 27 052 A1

(21) Aktenzeichen: P 37 27 052.4
(22) Anmeldetag: 13. 8. 87
(43) Offenlegungstag: 3. 3. 88

(30) Unionspriorität: (32) (33) (31)
21.08.86 JP P 61-196393

(71) Anmelder:
Minolta Camera K.K., Osaka, JP

(74) Vertreter:
Glawe, R., Dipl.-Ing. Dr.-Ing., 8000 München; Delfs, K., Dipl.-Ing., 2000 Hamburg; Moll, W., Dipl.-Phys. Dr.rer.nat., 8000 München; Mengdehl, U., Dipl.-Chem. Dr.rer.nat.; Niebuhr, H., Dipl.-Phys. Dr.phil.habil., Pat.-Anw., 2000 Hamburg

(72) Erfinder:
Murasaki, Sadanobu, Isehara, Kanagawa, JP

(54) Bilderzeugungsgerät

Zwischen einer Bilderzeugungseinheit und einer automatischen Entwicklereinheit (2), die angrenzend an die Einheit angeordnet ist, ist eine Zwischenkammer (33) mit einer Ablufteinrichtung (34) zum Transport des photoempfindlichen Materials (8) von der Bilderzeugungseinheit (1) zur Entwicklereinheit (2) vorgesehen. Selbst wenn Chemikaliendämpfe der in der Entwicklereinheit (2) aufgenommenen Lösungen zur Entwicklerbehandlung in die Zwischenkammer (33) durch eine Lücke in einer Auslaßöffnung (32) zur Zuführung des Materials (8) in die Entwicklereinheit (2) einströmen, wird verhindert, daß sie in die Bilderzeugungseinheit (1) einströmen, da sie zeitweilig in der Kammer (33) zurückgehalten und an einen Abluftkanal (35) ausgegeben werden. Da die Ablufteinrichtung (34) durch Abführen von Luft aus der Kammer (33) das Einströmen von Chemikaliendämpfen von der Entwicklereinheit (2) in die Zwischenkammer (33) bewirkt, wird von der Vorrichtung Luft aus der Bilderzeugungseinheit (1) in die Kammer (33) durch eine Lücke in der Einlaßöffnung (31) für das Zuführen des Materials in die Kammer (33) angesaugt, wodurch verhindert wird, daß Dämpfe aus der Kammer (33) in die Bilderzeugungseinheit (1) einströmen.

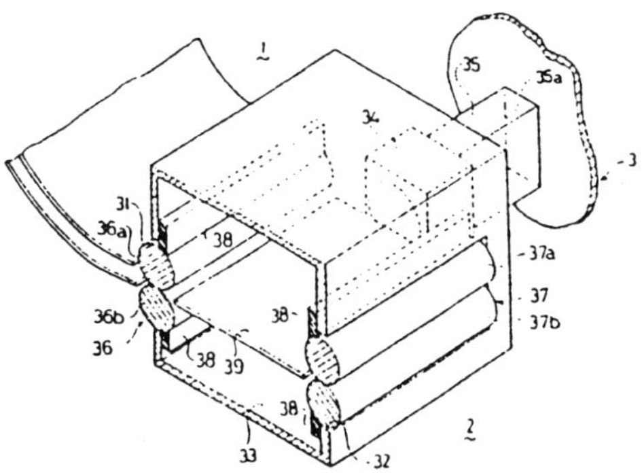

DE 37 27 052 A1

BUNDESDRUCKEREI 01. 88 708 869/538 14/60

Front page of a German unexamined patent application

(19) BUNDESREPUBLIK

DEUTSCHLAND

DEUTSCHES

PATENTAMT

(12) **Patentschrift**

(11) **DE 3544974 C2**

(51) Int. Cl. ⁴:

H 01 F 5/04

H 01 F 41/10
H 01 R 13/40
H 01 F 7/06

(21) Aktenzeichen: P 35 44 974.8-33
(22) Anmeldetag: 19. 12. 85
(43) Offenlegungstag: 25. 6. 87
(45) Veröffentlichungstag
der Patenterteilung: 23. 12. 87

DE 3544974 C2

Innerhalb von 3 Monaten nach Veröffentlichung der Erteilung kann Einspruch erhoben werden

(73) Patentinhaber:

AWECO Kunststofftechnik Gerätebau GmbH & Co
KG, 7995 Neukirch, DE

(74) Vertreter:

Eisele, E., Dipl.-Ing.; Otten, H., Dipl.-Ing. Dr.-Ing.,
Pat.-Anw., 7980 Ravensburg

(72) Erfinder:

Schrott, Hugo, 7995 Neukirch, DE

(56) Für die Beurteilung der Patentfähigkeit
in Betracht gezogene Druckschriften:

DE   33 16 456 A1
DE-GM   18 46 687
US   42 36 131
US   35 35 666
US   34 20 260
SIEBERT, Hanns-Peter: Physikalische Grundlagen
der Steckverbindertechnik. In: Der Elektroniker,
1981, H. 12, S. 17-22;

(54) Elektromagnet mit Steckanschluß

DE 3544974 C2

BUNDESDRUCKEREI   11. 87   708 152/407   80

Front page of a German examined patent application

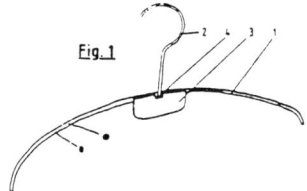

Fig. 1

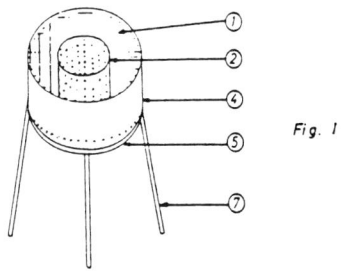

Fig. 1

(51) A 47 H – 7/02   (11) **36 21 070** (Z 1, T 5) A 1
(22) 24.06.86   (43) 07.01.88
(54) **Leiste zum Befestigen von Vorhängen**
(71) Thimm, Gerd, 4050 Mönchengladbach, DE; Hons, Arthur, 5170 Jülich, DE
(74) Cohausz, W., Dipl.-Ing.; Knauf, R., Dipl.-Ing.; Cohausz, H., Dipl.-Ing.; Werner, D., Dipl.-Ing.; Redies, B., Dipl.-Chem. Dr.rer.nat., Pat.-Anw., 4000 Düsseldorf
(72) Gleich Anmelder

(57) 1. Leiste zum verschieblichen Befestigen von Vorhängen und/oder Gardinen durch mindestens eine an der Leiste befestigte Schiene, in oder an der Gleitkörper oder Rollen laufen, die an dem oberen Rand des Vorhanges oder der Gardine befestigt sind, wobei eine Einrichtung mit zweifach umgelenkten, zur Seite geführten Schnüren vorgesehen ist, durch die der Vorhang oder die Gardine herablaßbar ist, **dadurch gekennzeichnet**, daß die einzelne(n) Schiene(n) (7) an den Schnüren (6) befestigt und gegenüber der unbeweglichen Leiste (1) absenkbar ist (sind).

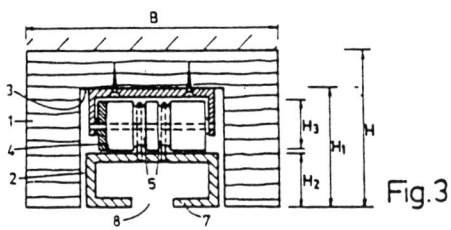

B

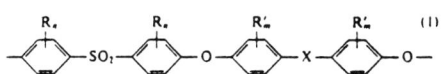

Fig.3

(51) A 47 J – 36/04   (11) **37 21 517** (Z 0, T 4) A 1
(22) 30.06.87   (43) 07.01.88
(30) 02.07.86 DE 3622087.6
(54) **Beschichtetes Kochgerät**
(71) BASF AG, 6700 Ludwigshafen, DE
(72) Koch, Jürgen, Dr., 6708 Neuhofen, DE; Heinz, Gerhard, Dr., 6719 Weisenheim, DE; Buchert, Hermann, Dr., 6702 Bad Dürkheim, DE
(51) B 29 C – 35/02   B 29 C – 69/00   B 05 D – 3/02
     B 05 D – 7/14   C 09 D – 3/49

(57) 1. Mit einem aromatischen Polysulfon beschichtetes Kochgerät, **dadurch gekennzeichnet**, daß das aromatische Polysulfon ein Copolymeres der allgemeinen Formel

$$R_a \quad\quad R_a \quad\quad R'_m \quad\quad R'_a \quad\quad (I)$$

$$-\bigcirc-SO_2-\bigcirc-O-\bigcirc-X-\bigcirc-O-$$

ist, wobei in der Polymerkette X zu 95 bis 50% SO_2- oder $C(CH_3)_2$-Brücken und zu 5 bis 50% S-Brücken sind, und wobei R und R' Halogen oder Alkyl mit 1 bis 4 Kohlenstoffatomen und n und m ganze Zahlen von 0 bis 4 bedeuten.

(51) A 47 J – 37/07   (11) **36 21 904** (Z 1, T 5) A 1
(22) 30.06.86   (43) 07.01.88
(54) **Holzkohlengrill mit Heißluftzone**
(71) Golf GmbH & Co KG, 4520 Melle, DE
(72) Goschke, Kurt, 4520 Melle, DE

(57) 1. Holzkohlengrill mit Heißluftzone mit einem Gehäuse aus Stahlblech bzw. Aluminium oder Nirosta mit einem innenliegenden gegenüber dem Gehäuse kleineren Glutkorb, um die Heißluftzone zu erzeugen.

| A 47 J – 47/01 | 37 21 518 | B 65 D – 83/04 |
| A 47 J – 47/01 | 37 21 519 | B 65 D – 83/04 |

(51) A 47 K – 1/14   (11) **36 21 715** (Z 2, T 10) A 1
(22) 28.06.86   (43) 07.01.88
(54) **Ablaufstopfen für ein Sanitärventil**
(71) Hans Grohe GmbH & Co KG, 7622 Schiltach, DE
(74) Ruff, M., Dipl.-Chem. Dr.rer.nat.; Beier, J., Dipl.-Ing.; Schöndorf, J., Dipl.-Phys., Pat.-Anw., 7000 Stuttgart
(72) Neugart, Dietmar, Ing.(grad.), 7233 Lauterbach, DE
(51) E 03 C – 1/22
(56) Fundstellen ermittelt.

(57) 1. Ablaufstopfen für ein Sanitärventil mit einem Ventilkelch (11, 51), wobei der Stopfen (19) einen Ventilteller (28, 44) mit einem in dem Ventilkelch (11, 51) angeordneten zylindrischen Ansatz (29, 41) und an dessen Unterseite mit einer Führungseinrichtung (20) aufweist, **dadurch gekennzeichnet**, daß der Durchmesser des Ventiltellers (28, 44) mindestens so groß ist wie der Außendurchmesser des Ventilkelchs (11, 51), der Ventilteller (28, 44) und sein Ansatz (29, 41) aus leichterem Material, insbesondere Kunststoff besteht, und an dem Ablaufstopfen (19) ein Gewicht (38) befestigt ist.

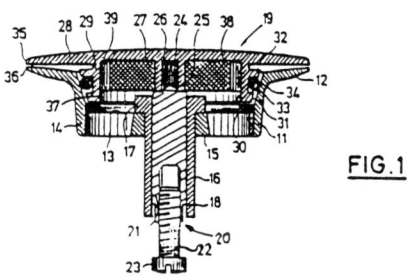

FIG.1

(51) A 47 K – 10/22   (11) **36 21 493** (Z 1, T 17) A 1
(22) 27.06.86   (43) 07.01.88
(54) **Rollenhalter, insbesondere für Toilettenpapier-Rollen**
(71) Erich Schumm GmbH, 7157 Murrhardt, DE
(74) Ruff, M., Dipl.-Chem. Dr.rer.nat.; Beier, J., Dipl.-Ing.; Schöndorf, J., Dipl.-Phys., Pat.-Anw., 7000 Stuttgart
(72) Kleemann, Karl Friedrich, 7157 Murrhardt, DE

(57) 1. Rollenhalter, insbesondere für Toilettenpapier-Rollen (11), mit einem Rollen-Gehäuse (2), in dem ein eine Lagerachse (8) bestimmender Aufnahmezapfen (7) für den Eingriff in eine Rollenkern-Öffnung (20) der jeweils in einer Rollen-Abzugsstellung befindlichen Rolle (11) sowie eine

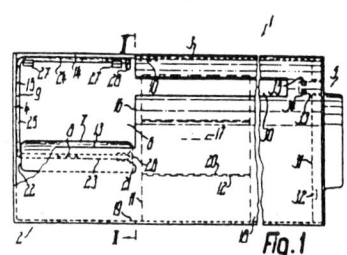

Fig.1

Classified entries for published German patent applications

The unification of Germany

The merger of the Federal Republic of Germany with the German Democratic Republic on the 3 October 1990 has not meant any changes to the documentation. This is because the pending East German applications at the time of the merger are continuing to be published as DD patents, or to be abstracted in the old East German gazette, the *Bekanntmachungen*. Applicants were given the option to continue a pending application through the West German system. Applications after the 3 October 1990 must go through the Munich office as (West) German applications, and the old East German material will eventually cease to be published once they have all been dealt with.

Legislation in the summer of 1991 means that the old East and West German patents have protection retrospectively extended to all of Germany. It had been provisionally decided to allow them protection only in the old jurisdictions.

SRIS and the German patents

SRIS keeps the three types of German specifications (OS, AS and PS) in separate sequences, but since a patent application always retains its original number it is easy to check for the most recent publication. In addition there are registers indicating to what stage an application reached.

None of the documents in the two supplementary number ranges for EPO or PCT specifications are held at SRIS.

Further reading

The unification of Germany and its effect on patents, T. Reimann and J. Feldges. *Patent World*, Dec 1990/Jan 1991. Issue 28, 31-35.

7 JAPANESE PATENT PUBLICATIONS

Numeration

The Japanese Patent Office allocates Japanese specifications numbers which have a prefix giving the number of years since the Emperor came to the throne, followed by a running number starting from 1 each year.

During 1988 there were numbers like 63 – 103582, since the Emperor had come to the throne in 1925, 63 years before). In 1989 the Emperor Hirohito died, which meant that after prefixing specification numbers with 64 for the first week, those for the rest of the year received the prefix 1. The running numbers, however, formed a single sequence for the entire year.

Identical numbers are used for different inventions when the applications are filed, when the unexamined applications 'kokai' are published, and when the examined patents are published. The numbers are used in the same way for utility models. Therefore the same numbers can be used up to six times for different applications.

Once they are granted, a running number is assigned to the patent or utility model (in separate series), with the patent numbers over 1,500,000 at present. These numbers are printed on an additional series of documents. They are rarely cited.

Numbers within the 500,001 onwards range of the published applications are used for translations of Patent Cooperation Treaty applications that designate Japan.

> **Identifying a Japanese patent number**
>
> The following suggestions may be helpful in trying to identify which sequence a citation belongs to.
>
> - Citations are usually to unexamined patent applications. *Chemical Abstracts*, for instance, usually cites these 'kokai' documents, and calls them by that name. In the absence of any other information
>
> - The documents bear (in Western characters) the designations 'A' for unexamined patent application, 'B2' for examined patent, and 'U' for unexamined utility model and 'Y2' for examined utility model.
>
> - Unexamined applications have only been published since 1970.
>
> - Examined documents are at present never numbered above 100,000, so higher numbers will clearly be for filings or unexamined applications.
>
> It is of course possible to look up the citation online to see if the IPC number or abstract sounds correct. The drawings in the 'kokai' and examined documents can also be compared.

Patenting procedure

Japan publishes the patent application, or 'kokai', 18 months after the priority date. The letter 'A' is prominently given at the top right of the front page. Applications filed before 1971 were only published once.

As in Germany the applicant has seven years from the date of filing before examination must be asked for. If it is acceptable then it is published a second time as an examined specification, or 'B2' on the front page.

The specification is published a third time when it is granted. About 250,000 applications are published annually, and about 60,000 examined specifications.

Specifications

Page 57 gives the front page of the unexamined application JP 63 - 219215. The date of publication, 12 September 1988, is given below the specification number. INID codes are used in the normal way. The drawings, unfortunately, are not given on the front page. Search reports are only printed on the examined documents.

Only claims with drawings are published of utility models at the application stage. They are sometimes referred to as 'Jitsuyo shinan koho' (their Japanese name).

Japan is unusual in publishing its patent documents in paperbacks containing a hundred or so documents, rather than as separate documents.

Both Derwent Publications and Inpadoc provide microfiche indexes and online databases for unexamined and examined patents, but neither have data on utility models, which are rarely cited. Utility models can be searched on the Patolis database (see page 87).

⑲ 日本国特許庁（ＪＰ）　　　　⑪ 特許出願公開

⑫ 公開特許公報（Ａ）　　昭63-219215

⑤Int.Cl.⁴　　　　識別記号　　　庁内整理番号　　　　㊸公開　昭和63年（1988）9月12日

H 03 G　3/10　　　　　　　　　A－7827－5J
H 03 F　1/02　　　　　　　　　　7827－5J
　　　　3/68　　　　　　　　　B－6658－5J　　審査請求　未請求　発明の数 1　（全3頁）

㊴発明の名称　　電力制御装置

　　　　　　　㉑特　　願　昭62－51913
　　　　　　　㉒出　　願　昭62(1987) 3月9日

㉔発 明 者　濱　田　　國　廣　　東京都国分寺市東恋ケ窪1丁目280番地　株式会社日立製
　　　　　　　　　　　　　　　　　作所中央研究所内

㉔発 明 者　塚　本　　信　夫　　東京都国分寺市東恋ケ窪1丁目280番地　株式会社日立製
　　　　　　　　　　　　　　　　　作所中央研究所内

㉔発 明 者　中　村　　　徹　　　東京都国分寺市東恋ケ窪1丁目280番地　株式会社日立製
　　　　　　　　　　　　　　　　　作所中央研究所内

㉚出 願 人　株式会社日立製作所　東京都千代田区神田駿河台4丁目6番地
㉔代 理 人　弁理士　小川　勝男　外1名

明　　細　　書

1. 発明の名称

　電力制御装置

2. 特許請求の範囲

1. 電力を電力分配回路で分配し、それらの電力
　を、並列に接続した電力増幅器でそれぞれ増幅
　し、しかる後にこれらの出力電力を合成回路で
　合成する電力合成型の電力増幅器において、上
　記並列接続した電力増幅器のバイアス電源を順
　次、若しくは、種々の組み合わせで、切りかえ
　て出力電力を可変する手段を具備してなること
　を特徴とする電力制御装置。

3. 発明の詳細な説明

〔産業上の利用分野〕

　本発明は、電力増幅器の出力電力の制御回路に
係り、特に、低い出力電力においても高い電力変
換効率（以後、単に「効率」と呼称する）の得ら
れる電力制御装置に関する。

〔従来の技術〕

　従来、移動通信用の無線機に使用される電力増

(1)

幅器は、第2図に示すように、電力増幅素子1－
4，2－2を縦続接続したものが多く用いられて
いる。この種の電力増幅器の出力電力に応じた効
率特性を第3図に示す。図示したごとく、出力電
力の高いときには、高い効率が得られているが、
出力電力が低いときには、効率が悪くなっていた。

〔発明が解決しようとする問題点〕

　従来知られている回路では、高出力電力のとき
最大効率が得られるように設計されているが、低
出力電力のときは、効率も低下するという問題が
あった。この原因は、電力増幅器が、通常最大出
力時に、最大効率が得られるように整合回路が設
計されるためである。しかしながら、たとえば、
自動車電話等に代表される移動通信機においては、
必ずしも常に最大出力電力で使用されるとはかぎ
らず、特に基地局近辺においては、低い出力電力
で使用されることが多い。従来の電力増幅器はこ
の点が考慮されていなかった。

　本発明の目的は、出力電力が低いときにも、高
い効率の得られる出力電力制御の手段を提供する

(2)

Front page of a Japanese unexamined patent application

The gazette

The gazette consists of the printed specifications themselves (with some additional material), namely the *Patent Application Gazette* and the *Patent Gazette*, and the *Utility Model Application Gazette* and the *Utility Model Gazette*.

Abstracts

Derwent Publications publishes *Japanese Patents Abstracts* (formerly *Japanese Patents Gazette*) in two series, for unexamined and examined, in both cases covering chemical patents only. It has been published since 1961.

Since 1977 the Japanese Patent Office has been publishing *Patent Abstracts of Japan*. This gives English abstracts plus drawings of most applications by Japanese nationals. It is published in several subject sections, with each issue giving the abstracts in numerical order.

SRIS and Japanese patents

SRIS will assume in photocopy orders, unless otherwise stated, that 'kokai' specifications are wanted.

None of the documents in the supplementary number range for PCT specifications are held by SRIS.

Granted Japanese patents (i.e. the third stage of publication) are not held by SRIS.

SRIS has a microfiche concordance which gives the subsequent numbers for each Japanese filing number with a single sequence for 1955-77 and thereafter annual sequences to 1987.

Further reading

The growth of Japanese science and technology, F. Narin and J.D. Frame. *Science*, Aug 1989. 245, 600-605 [Examines Japanese patenting policy in the United States].

The growth of Japanese Patent Office and its dissemination policy, A. Nakamura. *World Patent Information*, 1991. 13 (3), 125-138.

8   PATENT CLASSIFICATION

Why are patents classified?

The information content of patents is useless unless it can be retrieved reliably and quickly. Therefore there is a need for classification. However, it is important to remember that the primary purpose of patent classification is to help the patent examiners in their work.

Because of the nature of patent documents, classification schemes have been devised specially for them. The universe of knowledge, as far as patents are concerned, is all patentable subjects, a much wider field than is the case for most specialised classifications, yet the classification has to be extremely detailed. Hence patents classification schemes are bound to be massive. The most widely used is the International Patent Classification (IPC), which is used by the patent offices of over 70 countries. In addition, some countries use their own scheme, in particular Great Britain and the US. On the front page of patents (and elsewhere) the IPC classmark is shown at INID code 51; the domestic classification, if any, at code 52. The very widespread use of IPC is extremely important, since using it, searches can be carried out through the international patent literature, rather than having to search separately through each country's documents using a number of different schemes.

The International Patent Classification

General

The IPC is administered by the World Intellectual Property Organization. It is currently in its 5th edition, which has been in force since January 1990. (The 1st edition was in force from 1968-74; the 2nd edition from 1975-1979; the 3rd edition from 1980-1984 and the 4th from 1985-1989.) The edition in use is shown by a superscript number (Int. Cl.^{3} etc) on the front of patent documents. Because the scheme is international, revision is a complex and cumbersome matter, with a number of international committees of various levels of authority. The scheme comprises: the schedules, arranged by sections A-H; a catchword index; a guide; and a concordance with earlier editions. Sections A-H are as follows:

A: Human Necessities
B: Performing Operations, Transporting
C: Chemistry, Metallurgy
D: Textiles, Paper
E: Fixed Constructions
F: Mechanical Engineering, Lighting, Heating, Weapons
G: Physics
H: Electricity

These publications are available in printed form and are also to be found on the IPC:CLASS CD-Rom disc (see page 90).

Hierarchy

The scheme is hierarchical. The sample page on page 61 illustrates this. Each of the 8 Sections is broken down into Classes, e.g. class AOl: Agriculture, forestry etc; each class is broken down into Subclasses e.g. AOlB: Soil working in agriculture or forestry etc.; each subclass into Main groups e.g. AOlB 3/00: Ploughs with fixed plough-shares. Down to this level, the division is complete, i.e. at each level, the entire contents of that level are divided up into the level below. Also, down to this level, the hierarchy is reflected in the notation: it is clear that AOlB 3/00 is part of AOlB, which is part of AOl etc.

Each main group is divided into Sub-groups e.g. AOlB 3/02: Man-driven ploughs with fixed plough-shares. It is important to note that at this level the division is not necessarily complete, in other words there may be subjects which are not covered by any of the subgroups and are therefore classified at main group level. Also, at this level the hierarchy is not expressed in the notation, but in the dot structure, as shown on the sample page.

Each dot represents the next level in the hierarchy (which are called 1 dot subgroups, 2 dot subgroups etc). It is easy for the human eye to see and take in the hierarchy and realise that 3/62 is logically part of 3/60, which in turn is logically part of 3/58 etc but it is very important to remember when searching online that the dot structure is not built into the database index: searching for patents classified at subgroup A01B 3/58 will retrieve just that, not patents classified at e.g. 3/60. Most databases allow searching at subclass, main group or subgroup level without the need for truncation, other levels can be searched with the use of the appropriate truncation symbol.

The hierarchical structure has two important consequences. Firstly, all titles have to be read in the context of the higher terms. For example, the printed title at subgroup AOlB 1/16 says 'Tools for uprooting weeds', but this has to be read as 'Hand tools for uprooting weeds' because of the title of the main group, AOlB: Hand tools. Secondly, all notes (see below) will apply to all lower levels of the hierarchy.

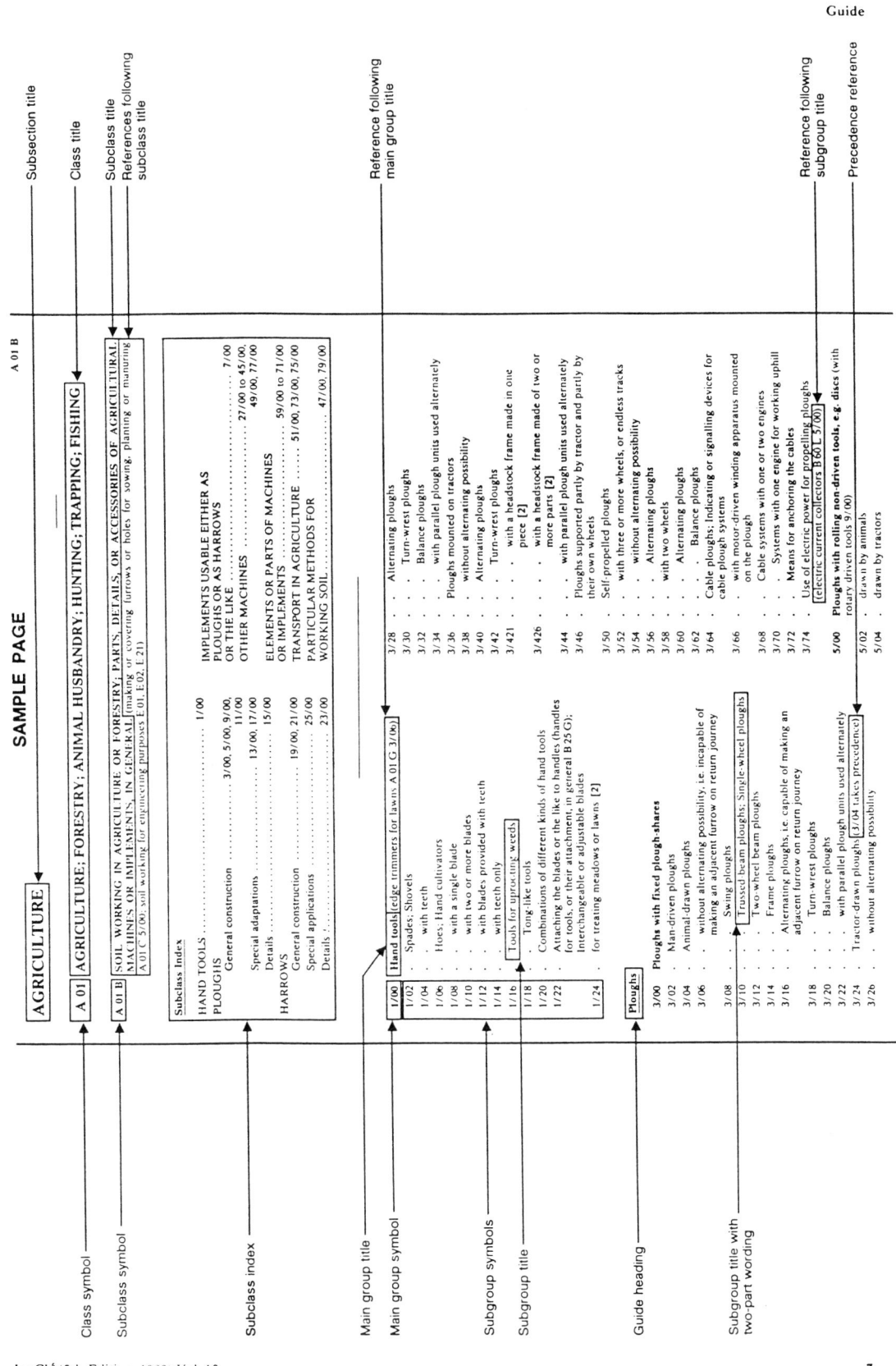

A page from the schedules of the International Patent Classification

Principles of classification

Patent documents can be classified by either of two principles. Documents relating to similar solutions to the same technical problem can be grouped together: this is known as classification by function (a valve for controlling the flow of fluids, for example, is essentially the same piece of equipment whether it is being used in a brewery or a sewage treatment plant). Alternatively, documents can be grouped according to the industrial art in which the problem arises, e.g. brewing technology: this is known as classification by application.

A patent classification scheme has to be based primarily on one of these principles if wholesale cross-classification is to be avoided. The IPC is based on the principle of classification by function, wherever possible. However, it does also include application classmarks, often, but not always, with a precedence note. (Chapter III of the *Guide* to the IPC addresses this problem.) This can lead to inconsistency between patent offices. To overcome this searchers should use both possible classmarks. For example, a searcher wanting documents on a gas burner for bakers' ovens designed to produce an even temperature distribution must look at A21B 001/28 (Bakers ovens heated by gaseous combustion products) and F23D (combustion apparatus using fluent fuel).

Problems in use

Even without the function/application problem, one is bound to get different classifications from different offices, for a number of reasons:

- The differing patent laws, e.g. with respect to non-obviousness, will influence the finding needs of examiners, and hence the classification policy of the office.

- The different language editions of the IPC: terms very rarely have exactly the same significance in different languages.

- Patent equivalents, though all describing the same invention, will not necessarily be expressed in the same way or stress the same points: a particular application might be stressed in one equivalent, for example.

It is also important to remember that some patent offices, such as Belgium and Finland, do not classify down to sub-group level.

The moral for searchers is: even if you think you have found the subgroup which is exactly what you want, searching only at that classmark is often not good enough.

Additional information and indexing

Under the convention governing the use of the IPC, patent offices are obliged to classify the inventive step information. They may also classify other matter disclosed in the patent (additional information), but this is discretionary. Since the introduction of the 4th edition they may also now use indexing in addition to the classification where specially indicated in the classification schedules. This additional information is called hybrid classification.

The difference between the two concepts is that classification attempts to locate the invention in one place (or a very limited number of places) that best fit the whole invention, whereas indexing codes try to identify various elements of information about the subject-matter (e.g. constituents of a mixture or uses of a product).

Both additional information and indexing codes are placed after the inventive classification and a double oblique stroke [//]. Indexing codes have the same type of notation, except that the slash [/] within the class is replaced by a colon [:]. An example is: B29C 65/08 // B29K 83:00, B29L 23:18 where B29C 65/08 is the classification symbol (joining preformed plastics parts using ultrasonics) and the remaining two codes are indexing codes (indicating, respectively, that the plastic is a silicon-containing polymer and that a hose is formed). Indexing codes are sometimes linked to a classifying symbol and in this case they are enclosed together in parentheses. An example is: (C08F 255/04, 214:06) where C08F 255/04 is the classification and C08F 214:06 is the indexing code.

The use of the indexing schedules is explained in the *Guide*.

Notes

Notes are an important part of IPC and essential to its correct use. There are a number of different types. One such type takes the form of references following the subclass title which give information on subjects not classed here. It is important to check for these before searching at group/subgroup level, because the note applies to the entire subclass. For example (see page 61), hand tools for making holes for planting will not be classified in A01B 1/00, because of the implicit note (given at A01B), Hand tools for making covering holes .... A01C 5/00.

Where references to other headings do not apply to the whole subclass, they will be given at the appropriate group or subgroup level (as for example at A01B 1/00 and 1/22). They will apply to the whole group, or to the particular subgroup, (including any dot subgroups within the subgroup) as appropriate.

The reference may send the user to a specific subgroup, or to another group, subclass or even class (such as the note at A01B). Or the note may be unspecific and leave the user to do more work.

This type of reference (away from the number to a more appropriate number) is the type of note most frequently used in IPC5. Another frequently encountered note is the precedence note, i.e. a note which indicates which of two possible classmarks should be used for a subject. There is an example at A01B 3/24 while no reciprocal note is given at A01B 3/04. A similar type of note is the last place rule, which is used particularly in the chemistry subject area.

It is important when searching for documents other than current ones to check whether the classification schedules for the subject have been changed between editions. Such changes are signalled in the text of the 5th edition: material new in the 5th edition is given in italics, while material introduced in earlier editions has [2] or [3] or [4] printed in bold beside the class mark. There is also a separate *Concordance*, showing alterations made between editions.

The *Guide* to the IPC contains much useful information and advice on how the scheme should be used, and should certainly be read before detailed use is made of the IPC. The Index is useful as a starting point in using the schedules, but it would be most unwise to search from it without referring to the Schedules and checking for the context, and for relevant notes. The Index appears not to be very consistent in the terms and concepts it includes.

Other schemes

Despite the widespread use of IPC, three prominent offices still use their own schemes.

ECLA

The *European Classification*, the scheme used by the *EPO*, is based on IPC (not necessarily the 4th edition) but with very many extra subdivisions. The scheme may be searched online on EDOC but is not printed on the patent documents. The text of the classification schedules is published as a loose-leaf manual and is also available online as ECLATX on Questel.

US Classification

The classification scheme used by the USPTO is important for searching US patents, as their assignment of IPC classmarks is sometimes unreliable. The scheme can be used to subject search US patents back to 1790 since all affected documents are reclassified when the schedules are amended. The scheme is technically very detailed, and somewhat user unfriendly. The classification schedules are available as a loose-leaf manual with a separate index and are also available on the CD-ROM system CASSIS issued by the United States Patent and Trademark Office (see page 90)

British Classification

In broad outline, the British classification key is very similar to IPC, in that it is divided into sections A-H. However, the similarity stops there. The scheme is very much more complex than IPC, and is very much geared to the needs of the Patent Office examiners. It is not available for online searching.

The classification schedules are republished periodically (currently every two years). The scheme is described in detail in three free pamphlets available from the Patent Office:

- *Structure of the classification key*

- *Notes on the use of the classification key*

- *Heading U1S Universal Indexing Schedule for use, application, utility and property*

The eight sections are divided into a number of divisions, which correspond to IPC class level. Abstracts are published weekly in subject groupings (see page 16). Lists are available from the Patent Office (file lists) giving the patent numbers of all patent documents classified at a particular term, or combination of terms.

In 1983, a new heading, U1S, was introduced, in response to criticism that revisions to the British scheme had made it increasingly less useful to industrial searchers. The Universal Indexing Schedule for use, application, utility and property is a set of indexing codes that can be applied to documents classified anywhere in the schedules. It is important to remember that U1S codes are intended to complement the classification, so if the use of an invention is clear from the classification, this information will not be repeated in the indexing. The heading consists of some 1320 terms in a categorised list, plus a separate schedule of utilities and properties of materials.

Further reading

International Patent Classification, 5th ed., 1989. World Intellectual Property Organization. Volume 10, Guide.

The Patent Office classification scheme, C. Oppenheim. *CIPA*, 1979. Supplement, 1-13 [Compares British classification with US and IPC].

Why 'hybrid' classification schemes ? D. Snow. *World Patent Information*, 1989. 11 (4), 200-207.

The International Patent Classification as a search tool, W. Vijvers. *World Patent Information*, 1990. 12 (1), 26-30.

This page intentionally left blank

9 PATENT INFORMATION RETRIEVAL SYSTEMS

It is often useful to retrieve technical information using one of the many abstracting services for specific technologies, some of which include patents in their coverage. But even for those services which include patents, the number of patenting countries included is often limited and there can be a considerable time delay in the appearance of the abstracts. There are however several abstracting services which cover patents thoroughly within their subject areas. *Chemical Abstracts Service*, for example, covers, in printed and online forms, patents from a large number of countries and is very thorough within its selection criteria.

The following two organisations produce services indexing and/or abstracting patents exclusively and cover an exceptionally large number of countries. The advantages given by these services include ease of access to a huge body of patent literature.

Derwent Publications This firm was established in the early 1950's for the prompt publication of abstracts of patent specifications. The abstracts were initially of British patents as at that time official abridgments appeared several months after the corresponding specifications, but were later expanded to provide abstracts of foreign patents in English. Initially on paper, the service was expanded in 1976 to provide remote online access to the database known as WPI (World Patents Index).

Abstracts are prepared to a standard pattern and convey the important features of the invention in about 120 words, often with a significant drawing (not available online), and a brief indication of the proposed use or advantage offered. The heading data includes an indication of whether the patent specification is a basic or an equivalent and the title of the invention shown on the specification is replaced by a more informative title (see page 68).

A   B   C   D E F

YEAR   SERIAL   WEEK

EAYT ★   Q51   87-158776/23   ★EP -225-096-A

Self-contained hydraulic bucket lifter for IC engine valve - has compliant diaphragm with number of circumferentially spaced accumulators absorbing flexure in reservoir pressure transients

G———   EATON CORP 14.11.85-US-798261 ————————————————————H

I————   *(10.06.87)●F01l-01/24* ————————————————————————J

K——14.11.86 as 308933 (1815SB)   (E)   EP-145445   EP-156260   J57126509———N

1.Jnl.Ref.R(DE FR GB IT)

O——The hydraulic lash adjuster (10) comprises a cup-shaped body having the closed end defining a face (18) adapted to contact an engine cam (16), the inner side of the closed end defining a precision guide surface. The adjuster includes a one-way valve slidably received against the guide surface and defining a force transmitting surface.

     The lash adjuster is operative to adjust and hold the distance of the transmitting surface from the cam face. A flexible wall is disposed to close the open end of the cup over the lash adjustment unit. The wall has the outer periphery sealed.

     ADVANTAGE - Minimises sideloading to minimise wear.

P——(16pp Dwg.No.1/8)

Q——N87-119177   L   M

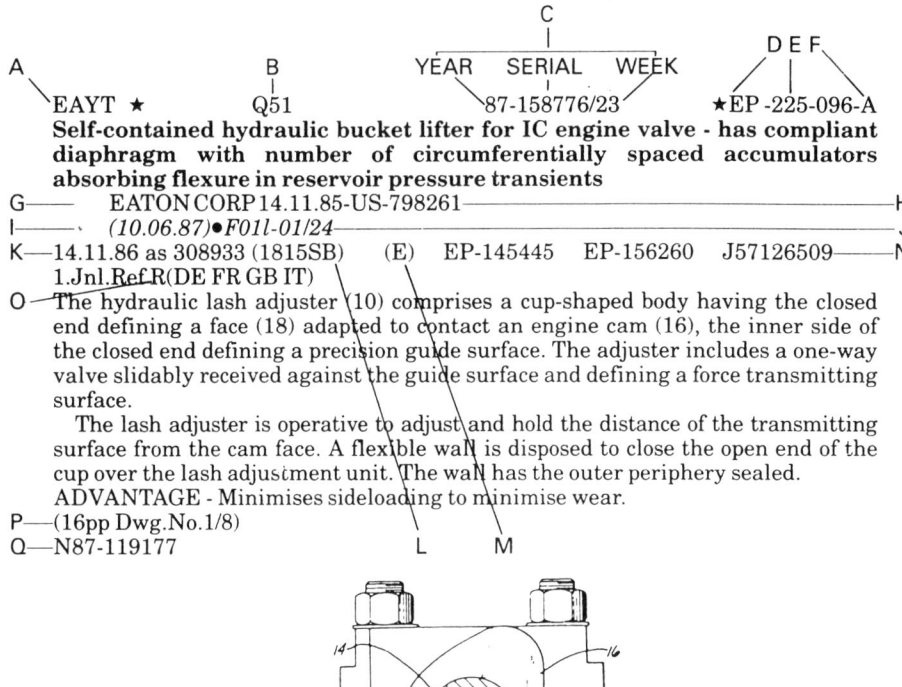

Key to Flagged Terms	Country Coverage and Codes
	Shown below are the 31 patent-issuing authorities covered by Derwent, together with their associated two character country codes.

Key to Flagged Terms

A – Patentee Code
B – Derwent class
C – Primary Accession No.
D – Basic (★) or Equivalent (=) indicator
E – Patent No.
F – Patent Status (A-1st, B-2nd, C-3rd Publication)
G – Patentee Name
H – Priorities (date, country, number)
I – Publication date of patent document
J – International Patent Classification
K – Local filing details
L – Abstractor's no. and Editor's initials
M – Language of Application
N – Citations
O – Designated States (N-National, R-Regional)
● – Basic Patent No. given here on equivalent headings

For Basics only:

P – No. of pages and Drawing Reference
Q – Secondary Accession No. (Microfilm auto-indexing No.)

Country Coverage and Codes

Country	Code	Country	Code
Australia	AU	Netherlands	NL
Austria	AT	Norway	NO
Belgium	BE	PCT	WO
Brazil	BR	Portugal	PT
Canada	CA	Romania	RO
China	CN	South Africa	ZA
Czechoslovakia	CS	South Korea	KR
Denmark	DK	Spain	ES
Europe	EP	Sweden	SE
Finland	FI	Switzerland	CH
France	FR	United Kingdom	GB
Germany, East	DD	U.S.A.	US
Germany, West	DE	U.S.S.R.	SU
Hungary	HU	Research	
Israel	IL	Disclosure	RD
Italy	IT	Int. Technology	
Japan	J5-J8	Disclosures	TP
Luxembourg	LU		

Typical Derwent Abstract

An important feature of the Derwent system is to group all patents relating to the same invention together under a single Derwent accession number. The first member of the patent family to appear on the Derwent database is the 'basic' and subsequent patents for the same invention are 'equivalents'. In general only one abstract is prepared for all the patents in the family.

The principal patent services currently offered by Derwent in printed form or in microform are *World Patents Abstracts, Chemical Patents Index, Electrical Patents Index, General Mechanical Patents Index*, and *World Patents Index*. They are all published weekly about 1-2 months after publication of the patent documents. Derwent covers the patents of about 30 countries patenting authorities as shown on page 70.

Chemical Patents Index (CPI). The *CPI* is a series of products for each of 12 subject sections: A polymers; B pharmaceuticals; C agricultural chemicals; D food and detergents; E general chemicals; F textiles and paper; G printing and photographic; H petroleum; J chemical engineering; K nucleonics and explosives; L ceramics and electrochemistry; and M metallurgy.

For each section Derwent publishes two series of *Alerting Abstract Bulletins* (arranged in country order and in classified order) containing 120 word abstracts as well as *Documentation Abstract Journals* containing longer abstracts. There are many other products within the CPI service including some which are based on the detailed Derwent subject indexing systems. CPI is available only as a subscription service.

Electrical Patents Index (EPI). In direct parallel with CPI, *Electrical Patents Index* provides *Alerting Abstract Bulletins* arranged in country order or in classified order for each of the EPI sections: S instrumentation; T computing and control; U semiconductors and electronic circuitry; V electronic components; W communications and X power engineering. There are no 'documentation abstracts' but a number of other products and detailed coding of the subject matter are provided.

General & Mechanical Patents Index. The *Index* is published as a series of abstract booklets for 4 subject areas viz. human necessities, performing operations, transport/construction, and mechanical engineering.

World Patents Abstracts. This provides specifications in all subject fields for the United States, Germany, Soviet Union, Belgium, United Kingdom, European Patent Office and Patent Cooperation Treaty and in the chemical field only for France (with South Africa), Japan and the Netherlands. Details of starting dates of these series are given on page 70.

World Patents Index. The *Index* is published in hard copy one week before the related abstracts and is in 4 sections, General, Mechanical, Electrical and Chemical. It covers about 30 issuing authorities and each issue contains the following indexes:

Derwent printed products coverage

Country	WPA	CPI	EPI	GMPI	WPI
AT Austria		1975-	1980-	1975-	1975-
AU Australia		1983-	1983-	1983-	1983-
BE Belgium	1955-	1970-	1980-	1975-	1974-
BR Brazil		1976-	1980-	1975-	1976-
CA Canada		1970-	1980-	1975-	1974-
CH Switzerland		1970-	1980-	1975-	1974-
CN China		1987-	1987-	1987-	1987-
CS Czechoslovakia		1975-	1980-	1975-	1975-
DD Germany former DDR		1970-	1980-	1975-	1974-
DE Germany (BRD)	1953	1970-	1980-	1975-	1974-
DK Denmark		1975-	1980-	1975-	1975-
EP European Patent Office	1978-	1978-	1980-	1978-	1978-
ES Spain		1983-	1983-	1983	1983-
FI Finland		1975-	1980-	1975-	1975-
FR France	1961-*	1970-	1980-	1975-	1974-
GB United Kingdom	1951-	1970-	1980-	1975-	1974-
HU Hungary		1975-	1980-	1975-	1975-
IL Israel		1975-	1980-	1975-	1975-
IT Italy		1978-	1980-	1978-	1978-
JP Japan	1962-*	1970-	1980-		1974-+
KR Korea (Rep of)		1987-	1987-	1987-	1987-
LU Luxembourg		1984-	1984-	1984-	1984-
NL Netherlands	1964-*	1970-	1980-		1974-
NO Norway		1975-	1980-	1975-	1975-
PT Portugal		1975-	1980-	1975-	1975-
RO Romania		1975-	1980-	1975-	1975-
SE Sweden		1970-	1980-	1975-	1974-
SU USSR	1961-	1970-	1980-	1975-	1974-
US United States	1971-	1970-	1980-	1975-	1974-
WO PCT	1978-	1978-	1980-	1978-	1978-
ZA South Africa	1961-*	1970-	1980-	1975-	1974-

Also coverage of International Technology Disclosures (1984-) and Research Disclosure (1978-) journals.

NOTES

W P A = World Patent Abstracts. Abstracts in individual country booklets.
  * = chemical coverage only.
C P I = Chemical Patents Index. Abstracts and detailed indexing.
E P I = Electrical Patents Index. Abstracts and indexing.
G M P I = General & Mechanical Patents Index. Abstracts series.
W P I = World Patent Index. Booklet and microfiche indexes.
  + = Chemical and electrical only.

Derwent printed products coverage

- Patentee Index, arranged in order of a four letter patentee code assigned by Derwent, e.g., British Petroleum is BRPE.

- IPC Index, arranged by International Patent Classification.

- Accession Number Index, a number assigned to each document on receipt by Derwent identifying patent families.

- Patent Number Index with accession number and patentee code.

There is also a weekly index covering all subjects:

- Priority Index, arranged by priority country and priority number and listing the patent numbers of each patent family.

The following indexes are all available on microfiches cumulated quarterly with a backfile 1974-78 and, chemical only, pre 1974.

- Patentee index. Similar to weekly index.

- IPC index. Similar to weekly index.

- Accession number index. Similar to weekly index.

- Patent Number Family Index, listing patent numbers of each family member for each patent issued.

- Priority Index, Similar to weekly index.

The index information may also be searched online. Further details are given in the next chapter.

Epidos-Inpadoc

The International Patent Documentation Center (Inpadoc) was established in Vienna in 1972 following an agreement between the World Intellectual Property Organization and the Austrian government. At the beginning of 1991 Inpadoc became part of the European Patent Office, and was absorbed into the new Epidos (European Patent Information and Documentation Systems).

Arrangements have been established with the principal patent offices of the world for the prompt supply of bibliographic data, often in machine readable form, and there are currently more than 50 subscribing patenting authorities. (See pages 72-73.)

The principal bibliographic elements of the patents documents are presented in the form of indexes on microfiche (as listed below) and similar information online. Unlike the Derwent service Inpadoc has no abstracts and the patent titles are given generally in their original language. However, titles in non-roman alphabets are given in English.

Indexes on Microfiche

Patent Family Service (PFS). PFS enables members of a patent family to be identified, the priority or basic documents being listed according to the country of filing or, in an extension of the service (PFS/INL), by the number of the priority document irrespective of the country of filing.

71

Inpadoc database coverage

Country	Dates				Language of title
	N D B	P C S	PAS / PAP	P I S	
AP ARIPO	84-11-01 -	84-11-01 -	84-11-01 -	84-11-01 -	English
AR Argentina	73-02-08 -	73-02-08 -	73-02-08 -		Spanish
AT Austria	69-12-10 -	72-03-10 -	69-12-10 -	69-12-10 -	German
AU Australia	73-01-18 -	73-01-18 -	73-01-18 -	73-01-18 -	English
BB Barbados	PCT only 85-09-26 -				English
BE Belgium	64-05-15 -	64-05-15 -	70-06-30 - 74-05-02; 79-07-02 -	70-07-02 - 74-05-02; 79-07-02-	French/Dutch
BG Bulgaria	73-02-15 -	73-02-15 -	77-10-12 -	77-10-12 -	English
BR Brazil	73-01-02 -	73-05-10 -	73-01-02 -	73-01-02 -	Portuguese
CA Canada	70-08-04 -	70-08-04 -	74-01-01 -	74-01-01 -	English/French
CH Switzerland	69-03-31 -	69-03-31 -	74-02-28 -	78-01-13	German/French /Italian
CN China	85-01-01 -	85-01-01 -	85-01-01 -	85-01-01 -	English
CS Czechoslovakia	73-02-22 -	73-02-22 -	77-10-31 -	77-10-31 -	English
CU Cuba	74-02-13 -	74-02-13 -	77-09-08 -	77-09-08 -	Spanish
CY Cyprus	75-03-01 -	75-03-01 -	75-03-01 -		English
DD Germany (DDR)	73-01-05 -	73-01-05 -	73-01-05 -	73-01-05 -	German
DE Germany (FDR)	67-01-05 -	67-01-05 -	71-02-18 -	73-01-04 -	German
DK Denmark	68-04-29 -	68-04-29 -	68-04-29 -	68-04-29 -	Danish
EG Egypt	76-01-31 -	76-01-31 -	76-01-31 -	76-01-31 -	English
EP European Patent Office	78-12-20 -	78-12-20 -	78-12-20 -	78-12-20 -	English
ES Spain	68-01-01 -	68-01-01 -	68-01-01 -		Spanish
FI Finland	68-05-31 -	68-05-31 -	68-05-31 -	68-05-31 -	Swedish
FR France	68-11-29 -	68-11-29 -	69-08-29 -	76-01-02 -	French
GB United Kingdom	69-04-30 -	69-04-30 -	73-02-14 -	83-01-06 -	English
GR Greece	77-07-04 -	77-11-05 -	77-07-04 -	77-07-04 -	English
HK Hong Kong	76-03-05 -	76-03-05 -	76-03-05 -	76-03-05 -	English
HU Hungary	73-01-29 -	73-01-29 -	77-09-28 -	77-09-28 -	English
IE Ireland	73-01-10 -	73-01-10 -	73-01-10 -		English
IL Israel	68-01-25 -	68-01-25 -	68-01-25 -		English
IN India	75-08-02 -	75-08-02 -	75-08-02 -	75-08-02 -	English
IT Italy	73-01-25 -	73-01-25 -	73-01-25 -	73-01-25 -	English
JP Japan	73-04-02 -	73-04-02 -	76-01-05 -	76-01-05 -	Romanized Japanese/ English
KE Kenya	75-07-11 -	75-07-11 -	75-07-11 -	75-07-11 -	English
KP Korea (DPR)	PCT only 79-04-05 -				English
KR Korea (Rep of)	78-02-28 -	78-03-04 -	78-02-28 -	78-02-28 -	English
LI Liechtenstein	EPO only				English
LK Sri Lanka	PCT only 82-09-16 -				English
LU Luxembourg	45-06-02 -	60-07-30 -	79-03-06 -	79-03-06 -	French/German
MC Monaco	75-10-10 -	75-10-10 -	75-10-10 -	75-10-10 -	French
MG Madagascar	PCT only 79-01-25 -				English

Inpadoc database coverage

	Country	NDB	PCS	PAS / PAP	PIS	Language of title
MN	Mongolia	72-11-20 -	72-11-20 -	78-12-25 -	77-02-05 -	English
MW	Malawi	73-05-09 -	73-05-09 -	73-05-09 -		English
MX	Mexico	81-01-01 -	81-01-01 -	81-01-01 -	81-01-01 -	Spanish
MY	Malaysia	53-12-31 -	53-12-31 -	53-12-31 -	53-12-31 -	English
NL	Netherlands	64-03-10 -	64-07-06 -	75-01-06 -		Dutch
NO	Norway	68-07-01 -	68-07-01 -	68-07-01 -	68-07-01 -	Norwegian
NZ	New Zealand	78-03-06 -	78-03-06 -	78-03-06 -	78-03-06 -	English
OA	OAPI	PCT only 78-12-21 -				English
PH	Philippines	75-07-03 -	75-07-03 -	75-07-03 -	75-07-03 -	English
PL	Poland	73-02-28 -	73-02-28 -	77-08-31 -	77-08-31 -	English/Polish
PT	Portugal	76-01-01 -	76-05-01 -	76-01-01 -	76-01-01 -	English/French /German
RO	Romania	73-01-20 -	73-01-20 -	77-10-15 -	77-10-15 -	French
SD	Sudan	PCT only 84-12-06 -				English
SE	Sweden	68-04-01 -	68-04-01 -	68-04-01 -	68-04-01 -	Swedish
SG	Singapore	83-02-25 -		83-02-25 -	83-02-25 -	English
SU	USSR	72-12-07 -	72-12-07 -	77-07-05 -	77-07-05 -	English
TR	Turkey	73-01-01 -	77-01-01 -	73-01-01 -		English
US	United States	68-01-02 -	68-01-02 -	68-01-02 -	68-01-02 -	English
VN	Vietnam	85-01-01 -	85-01-01 -	85-01-01 -	85-01-01 -	English
WO	PCT	78-10-19 -	78-10-19 -	78-10-19 -	78-10-19 -	English
YU	Yugoslavia	73-02-28 -	73-02-28 -	73-02-28 -	73-02-28 -	English/ Serbo-Croat
ZA	South Africa	71-01-27 -	71-01-27 -	71-01-27 -	71-01-27 -	English
ZM	Zambia	68-10-16 -	68-10-16 -	68-10-16 -		English
ZW	Zimbabwe	80-09-03 -	80-09-03 -	80-09-03 -	80-09-03 -	English

Patent Classification Service (PCS). The entries are listed in order of IPC and within each class in order of country of publication, publication date and kind of document. See the example on page 75.

Patent Applicant Service (PAS). The entries are grouped in alphabetical order of the name of applicant and within each group by IPC. See the example on page 76. An important extension to this service (PAP) lists the entries under each applicant's name in order of the priority number of the documents. This means that all the patent family members applied for by an applicant are grouped together.

Patent Inventor Service (PIS). The entries are arranged in alphabetical order of inventor's name. This service is especially useful for updating information on the work of a particular person whose line of research is already known to be of interest. It can also indicate development trends and changes in affiliation of key personnel. See the example on page 76.

Numerical Data Base (NDB). All the patent documents published by each country are listed in numerical order thus facilitating the tracing of stages of publication and of documents published in random numerical order. See the example on page 75.

All of the above services are published quarterly and cumulated into 5 year sequences. The back-file 1968-72, which was compiled retrospectively, contains data from eighteen countries. A list of the country coverage and commencement dates is given on pages 72-73.

Inpadoc Patent Gazette (IPG). This is a rapidly published, weekly service covering newly issued patent documents and contains the data used to compile the cumulated services. It is divided into four parts, Selected Classification Service (SCS), Selected Applicant Service (SAS), Selected Inventor Service (SIS) and Selected Numerical Service (SNS) corresponding to cumulated services PCS, PAS, PIS and NDB respectively. Equivalents are shown on all four of the weekly services.

Concordance, application number to publication number (CON). The identification of the published specification when only the application number is known is a frequently occurring problem. The official gazettes of the country may carry a concordance relating to the publications in that issue but there are few useful cumulated tables apart from those compiled by INPI for French patents and Japio for United States and Japan.

The CON service, relating application number to publication number and updated annually, is at present available for 32 countries, some back to 1968.

```
N D B  NUMERICAL DATA BASE              MICROFICHE   OCT/1984   PRODUCED:   84-10-01   PAGE: 52 623
PUB. COUNTRY : UNITED KINGDOM                                              FICHE NO: 279  FRAME: L13
CC DOC.NO KD PUB**DAT APP**DAT KA APPLIC. NO.  CC PR.**DAT KP PRIORITY NO.   IPC          APPLICANT       INVENTOR          TITLE

GB 2134526 A1 84-08-15 84-01-31 A 84 8402493   JP 83-01-31 A 83   14980     C07J 1/00    * SHIONOGI SEIY  SUSUMU * KAMATA   ALDOSTERONE-ANT
GB                                                                          A61K 31/565  AKU KABUSHIKI    TAKEAKI * MATSU   AGONISTIC STER
GB                                                                          A61K 31/585  KAISHA           I                 OIDS
GB                                                                          C07J 21/00                    NOBUHIRO * HAGA
GB 2134527 A1 84-08-15 84-02-03 A 84 8402912   GB 83-02-04 A 83   8303111   C08J 9/14    H R * SMITH      DEVENDRA PAL *    CROSS-LINKED FO
GB                                                                                                        SINGH             AMED THERMOPLA
GB                                                                                                                          STICS

                    ^^^^^^^^^^^                            ^^^^^^^^^^^^^^^^^^
                    Application                                Priority
                       data                                      data
```

```
P C S  PATENT CLASSIFICATION SERVICE    MICROFICHE   JUL/1984   PRODUCED:   84-07-01   PAGE: 2 238
                                                                           FICHE NO: 15  FRAME: F06
I P C        CC PUB**DAT KD DOC.NO IPC (ALL)  CC PR.**DAT KD PRIORITY NO.   APPLICANT        TITLE

A41H 37/10   KR 84-03-27 B1 8400363 A41H 37/10 KR 81-02-07 A 81  8100387    NPI NEW PRODUCT  DEVICE FOR THE FASTENING OF BUTTONS TO
                                                                            S INVESTMENT A   PIECES OF GARMENT
                                                                            B
             SE 84-04-16 B  432694  A41H 37/10 SE 78-03-16 A 78  7803025    BENGT PETERSSO   FORFARANDE FOR ANBRINGNING AV FOREMAL S
                                    A44B  1/18  SE 78-11-17 A 78  7811869    N NEW PRODUCTS   ASOM KNAPPAR PA TEXTILMATERIAL
                                               SE 79-01-18 A 79  7900444    INVESTMENT AB
A41H 41/00   SU 84-02-29 A1 1076069 A41H 41/00 SU 81-12-24 A 81  3370910    TS PK T BYURO L  APPARATUS FOR DRYING FUR SKINS OF TUBUL
                                    C14B 15/06                              EGKOJ PROMYSHL   AR FORM
                                    F26B 15/06                              ENNOSTI

                    ^^^^^^^^^^^                            ^^^^^^^^^^^^^^^^^^
                    Publication                               Priority
                       data                                     data
```

Inpadoc microfiche - Numerical Data Base (top) and Patent Classification Service (below)

P A S PATENT APPLICANT SERVICE   MICROFICHE   OCT/1984   83-12-31 - 84-09-28   PRODUCED: 84-10-01   PAGE: 25 937
FICHE NO: 131 FRAME: L08

APPLICANT	CC PUB**DAT KD DOC.NO	CC PR.**DAT PRIORITY NO.	I P C	INVENTOR	TITLE
LEIFKE VOLKE	DD 84-03-28 A1 208266	DD 81-12-07 81 235486	H01H 50/54	LEIFKE,VOLKER,DD	KONTACT KASSETTENSYSTEM F UER SCHALTGERAETE
LEIGH FLEXIBLE STRUCTURES	GB 83-01-19 B2 2042988	GB 78-12-11 78 7847948	B63B 43/10		FLEXIBLE BARGES
	GB 82-11-10 B2 2037350	GB 78-12-11 78 7847947	E02B 7/00		FLOOD BARRIER
	CA 84-07-31 A1 1171673	GB 80-10-27 80 8034511	E02B 7/02	HARDACRE, NICHOLAS P	FLEXIBLE BARRIERS
			E02B 3/10		

^^^^^^^^^^^   ^^^^^^^^^^^
Publication   Priority
   data   data

P I S PATENT INVENTOR SERVICE   MICROFICHE   OCT/1984   83-12-31 - 84-09-28   PRODUCED: 84-10-01   PAGE: 13 475
FICHE NO: 68 FRAME: J07

INVENTOR	CC PUB**DAT KD DOC.NO	CC PR.**DAT PRIORITY NO.	I P C	APPLICANT	TITLE
FIELDSTEEL A HOWARD	US 84-08-07 A 4464470	US 81-02-25 81 237917	C12N 1/20	SRI INTERNATIONAL	REPLICATION OF VIRULENT TREPONEMA PALLIDUM IN T ISSUE CULTURE
			C12N 5/00		
			C12N 1/00		
			C12N 1/36		
			C12N 1/04		
FIELK INGRID	RO 84-02-21 B 83325	RO 79-09-01 79 98567	A01N 25/02	ADMINISTRATIA PARCURI LOR SI STRAZILOR,RO	COMPOSITION POUR LA STIM ULATION DE LA RIZOGENES E
FIELOW ROBERT L	ZA 83-11-30 A 831424	US 82-03-02 82 354125	B62D	FRUEHAUF CORP	MEANS FOR REDUCING VEHIC LE DRAG

^^^^^^^^^^^   ^^^^^^^^^^^
Publication   Priority
   data   data

Inpadoc microfiche - Patent Applicant Service (top) and Patent Inventor Service (below)

Patent Register Service (PRS). This service provides information on the legal status of patent applications and currently covers many European countries, United States, EPO and PCT, the details being taken from the official gazette of each patenting authority. Against each patent number there is a code with a heading in the language of the issuing country and reported changes in legal status, e.g. granted, assigned, expired, withdrawn, etc.

This page intentionally left blank

10 PATENTS ONLINE AND COMPACT DISC SEARCHING

The first part of this chapter discusses ways of searching online or on compact disc, and the problems in devising a satisfactory search strategy. The second part consists of an annotated list of patents databases, both those held online and on compact disc, with a list of the online files arranged by host.

Searching databases always involves the user and a computer and keyboard. Online searching means that there is a telecommunications link to the database, typically held on a mainframe computer of the host (an organization which has set it up on their mainframe). The results can be received on the screen and can be printed 'online', that is, right away on the user's own printer, or 'offline', when the results are sent to the user by post.

Compact disc searching means that the compact discs are stored in a drive next to the user's computer, so that there is no need to use a host. Printing will be done on the user's own printer.

There are three main advantages in searching online, rather than manually, for patents information.

- It is very much quicker: a search that could involve hunting for hours through fiche or abstracting journals can be done in minutes online.

- You may not have access to all the relevant manual aids such as Inpadoc microfiches.

- You can carry out searches, particularly subject searches, which would be very cumbersome, or impossible, to do manually.

A particular advantage is the ability to easily combine search requirements: to find patents by a company on a certain subject, or over a certain time-span, for example.

However, there are three major drawbacks:

- online searching is expensive

- date coverage does not extend far enough back for many purposes

- drawings are not (yet) available

What's available online?

There are three major hosts as far as patents databases are concerned. Dialog, Orbit and Questel, each with a family of relevant databases. Other hosts with patents databases include STN, Mead Data Central and the Spanish Patent Office. There is some duplication between hosts, but Questel and Orbit each have important exclusive databases.

When choosing a host you will wish to consider a number of factors:

- The range of databases offered, both of patents and related non-patents files.

- The pricing structure of the host (in particular if it is a subscription service).

- What special features are available, such as cross-file searching or online statistical analysis (cross-file searching is particularly important for patents, firstly in that the search may need, for legal reasons, to be as complete as possible, and secondly in that the various bits of information needed are often not all available in the one database).

- What help is available from the hosts, such as documentation, training courses, online help and error messages.

- The availability of a document-ordering service (both Dialog and Orbit allow the online ordering of patent photocopies from the British Library's Patent Express service).

- Your own preference with regard to command language, formats etc.

Which database?

A very substantial amount of patents information is available online, covering 56 countries, over various timespans, and including bibliographic data, abstracts, and administrative and status information. However, in most cases this information is not all available in one database. In choosing which database, or databases, to use for any particular search, you will need to bear in mind the following points:

- The type of information needed, and the purpose for which it is needed. A subject search might be a state of the art search, a general technological enquiry, an infringement search, or simply to obtain the details of known existing patents. Is administrative or legal status data needed? Are correct bibliographic details needed? Is a family search needed to track down all family members, or is what is sought an English language equivalent?

- What information is already known: what companies are working in the field? the date or classmark of known relevant patents?

- The geographical coverage needed: GB only (plus EP), major countries, worldwide?

- The date span of coverage needed?

- How up to date the search results must be.

Name searching

Name searching – by applicant (or assignee, for US patents) or inventor can easily be carried out online, though it is not necessarily as straightforward as might be wished.

The Inpadoc database standardises names but the practice does not always match up to the theory: it may take some while for new variants to be recognised as such, and of course typographical errors can never be totally eliminated. Corporate names such as those of universities can cause particular problems. Online, of course, the problem of order of elements in a corporate name can easily be dealt with by searching for significant words in the name irrespective of the order in which they appear:

'Oxford and (univ or university)', for example, will retrieve University of Oxford, Univ Oxford, Oxford Univ or even Chancellor of Oxford Univ. Another technique is to search the index to the database (neighbour / expand / ..ind) which enables one to see and retrieve variants of a name.

Applicant names can also be searched for in Derwent's WPI database, using keywords or phrases, or by the Derwent company code, which is usually more reliable but means that you may also pick up patents from related companies, as Derwent will normally assign the same code to all the companies in a group. Inventor names can also be searched on WPI for records since 1978. With inventors, you need to be particularly careful over how forenames or initials are treated by the different hosts. Inventor address information is not available on either WPI or Inpadoc, but is available for European and US patents, on the databases EPAT and USPA respectively. EPAT also includes licensee information.

Number searching

On both Inpadoc and WPI you can search by priority or application number; however, searching by priority number on Inpadoc – i.e. to create a patent family – attracts a surcharge. By contrast, family information is easily available from WPI, though it covers fewer countries and, if there is any complication over priority dates, may not always be as complete as that which is available from Inpadoc. Derwent records online include the patent numbers and associated application data for all patents which Derwent regards as equivalents (i.e. including non-convention equivalents), so more or less complete family information is instantly and easily available. On WPI on Orbit, use of the 'Fpat' command links records sharing any priority date. Family information is also available from the EPO file, available on Questel as EDOC. Family searching on this file also attracts a surcharge.

Legal status

Inpadoc's Patent Register service (see page 77) is also available online, where it is updated weekly, and with the added advantage that all the status information is translated into English.

There is a certain amount of other legal status data available online. American patents are particularly well covered for status data by the files Claims/Reassignments, Litalert and Patent Status, which all cover different aspects of post-issue actions or litigation. Status data for EP patents is included in the EPAT file, and for German patents in PATDPA. Legal action over French patents is covered by Jurinpi. Lexis contains a number of databases reporting legal action over British and US patents. The status of pharmaceutical patents is covered in Drug Patents International.

Subject searching

There are a number of ways to approach subject searching for patents. It is possible to use keywords, or classification (the International Patent Classification, or the EPO or US classification), or the Derwent classification. The British classification is at present not searchable online. Often a combination of keyword and classification can produce the best results.

IPC can be used to search online, on Inpadoc, WPI, and on the American patents files. The EPO's modification of IPC: ECLA, can also be searched online on the EDOC file, but only to produce file lists (i.e. lists of patent numbers classified at any particular classmark). Because the database does not include any text, one cannot use keywords to modify a search, nor print titles to check relevance of patents retrieved. The only reason for choosing to use this file for subject searching is its date coverage: it goes back further than any other for non-US patents. British patents, for example, are covered from 1909, French from 1902 and US from 1920.

The US Classification - a more reliable method of searching US patents than IPC, due to the USPTO's method of assigning IPC codes - is also available for searching online. All US patents are covered, back to 1790. Whenever the US Classification is amended, all affected patents are reclassified, so there is no need to check through earlier editions of the classification for relevant codes.

The Derwent classification can be used to search in WPI: it can be useful particularly in conjunction with keywords, to narrow the scope down to a broad subject area, e.g. pharmaceuticals, which would be impossible to do with IPC. WPI is certainly the most useful file for subject searching online. The earliest records on the file date from 1963, but these are chemical patents only; general and mechanical patents are covered only from 1974 onwards. Because of the different coverage, in detail and in date, of each country and each subject area, it can be very difficult to determine precisely the scope of any particular search results. However, from 1974 onwards coverage is good enough for most purposes, and the subject searching possibilities are excellent, allowing searching by keywords, from the abstract or restricted to the title, either on their own or in combination with IPC or Derwent classification codes. Using these methods, a search can be made as narrow or as broad in scope as is wanted.

Keyword searching is also possible on the US files which include the text of the abstract and of one or more of the claims, USPA and Claims/Patents. A full text version of US patents is also available online on the Lexpat file, mounted by Mead Data Central.

For chemicals searching, a number of specialised services are available, such as the Derwent specialised coding system, available only to subscribers, and, for US patents, the Claims Uniterm and Claims Comprehensive files.

There are also a number of subject-specialised files, either confined to patents, such as Apipat (petroleum-related patents) or which include patents in their coverage. Most of the latter are the online versions of hard copy products.

Citation searching

Many patents are now published with their search report. This makes it possible to trace back the development of a particular technology. Using online methods, it is also possible to carry out reverse, or forward citation searching. For example, if a relevant patent is known in a particular subject area, it is possible to search online to see if it has been cited against any later applications, thus tracing forward the development of a technology. This can at present be done only against US, EP, PCT, French and German patents (ie. the cited patent can be of any country, but the later citing patent can only be one of those five). The databases which include citation information are USPA, Claims/Citation, EPAT, WPI (for EP and PCT patents only), FPAT and PATDPA.

Compact discs

Compact discs (CD-ROMs) are increasingly being used in patent information. This is in two areas.

The first is as a means of storing the complete text (including drawings) of a specification, in which online as yet does not compete.

The second is in providing databases for searching, which are mostly, but not entirely, bibliographic. An example is the CASSIS database. Non-bibliographic databases include the International Patent Classification and the US classification.

New products are appearing at a rapid rate, but most of the activity is presently confined to American or European Patent Office data.

The chief advantages of searching on CD-ROMs rather than online are:

- They are simpler to use

- They are cheaper to use (provided a large amount of searching is carried out)

- Printing or downloading is available

The chief disadvantages are:

- They tend to be less current than online (there are exceptions)

- Only one person can use them at a time (unless it is on a network)

- They tend to be less flexible in their searching capabilities

- They are nearly all restricted to one country or patenting system

- Coverage is usually for only a few months' or years' of patents, although the discs can be changed to extend the search

The currency of the information is clearly affected by the frequency of updating the CD-ROMs. Six or 12 times annually are common.

The very large capacity of online databases means that CD-ROMs are unlikely to ever completely replace online searching. They do however provide a quick, easy and cheap alternative for many searches.

A list of CD-ROM databases follows that of online databases, at the end of this chapter.

Further reading

Online patents, trade marks & service marks databases, J. Sibley. London: Aslib, 1991.

Possibilities and limitations of online searching, F. Torfs and G. Ashley. *Patent World*, July 1990. Issue 24, 24-28.

Usefulness of the hybrid systems in computerized searches, M. Nishikawa. *World Patent Information*, 1990. 12 (4), 212-215.

Patent information on CD-ROM, M. Lobeck. *World Patent Information*, 1990. 12 (4), 200-211.

On-line data bases for chemical patent searches, M. Herz. *World Patent Information*, 1990. 12 (3), 119-124.

CD-ROM: a new means of searching and storing patent information, E. Hearle. *World Patent Information*, 1991. 13 (3), 139-142.

Effective strategies for searching existing patent rights, J. van der Drift. *World Patent Information*, 191. 13 (2), 67-71.

Databases

The annotated list of databases is restricted to those on commercial hosts available in the UK, and does not claim to be complete.

The list of hosts given immediately below is restricted to those offering databases given in the annotated list.

Hosts and the databases they offer

BRS:
Maxwell Online Ltd.,
Achilles House,
Western Avenue,
London E3 OUA,
United Kingdom.
Tel 081 992 3456
Fax 081 993 7335
  Merck Index
  NTIS
  PATDATA

Data-Star:
Data-Star,
Plaza Suite,
114 Jermyn Street,
London SW1 6HJ,
United Kingdom.
Tel 071 930 5503
  CA Search
  Current Patent Evaluations
  NTIS

Dialog:
Dialog Europe,
PO Box 188,
Oxford OX1 5AX,
United Kingdom.
Tel 0865 730 275
  Apipat
  CA Search
  Chinapats
  Claims
  Inpadoc
  Merck Index
  NTIS
  World Patents Index

ESA-IRS:
IRS/DIALTECH,
Science Reference
and Information Service,
25 Southampton Buildings
London WC2A 1AW.
Tel 071 323 7951
Fax 071 323 7954
  CA Search
  ITALPAT
  Space Patents

Mead Data Central
Mead Data Central Inc.,
P.O. Box 933,
Dayton,
Ohio 45401,
United States.
Tel 513 865 6800
Fax 513 865 6909
  LEXIS (Produced by Butterworth)
  LEXPAT
  Tech-line

Orbit:
Maxwell Online Ltd.,
Achilles House,
Western Avenue,
London E3 OUA,
United Kingdom.
Tel 081 992 3456
Fax 081 993 7335
  Apipat (within World Patents Index)
  CA Search
  Chinapats
  Claims
  Current Patents Evaluations
  Drug Patents International
  Inpadoc
  JAPIO
  Litalert
  Merck Index
  NTIS
  PAST
  USCLASS
  USPA
  World Patents Index

Questel:
Questel,
83-85 Boulevard Vincent Auriol,
F-75013 Paris,
France.
Tel 1 45 82 64 64
Fax 1 45 85 43 08
  CA Search
  ECLATX
  EDOC
  EPAT
  FPAT
  Jurinpi
  Pharmsearch
  TRANSIN
  World Patents Index

STN:
STN-Karlsruhe,
c/o Fachinformationszentrum,
Energie, Physik, Mathematik,
GmbH,
Postfach 2465,
D-7500 Karlsruhe 1,
Germany.
Tel 7247 82 95 66
  Apipat
  CA Search
  Claims (IFIClaims)
  Inpadoc
  NTIS
  PATDPA
  PATOS

General databases

Inpadoc. Produced by Epidos (formerly Inpadoc) of the European Patent Office. Contains over 17 million records, from 56 patenting authorities. Consists only of bibliographic data: patent applicant, priority details, IPC, title, inventor (usually). Legal status data is also available for 10 major countries and for European Patent Office and Patent Cooperation Treaty publications.
Available on Orbit, Dialog and STN, with the structure of the file differing on the different hosts.

On Orbit, the legal status file is mounted separately and is known as Legstat. On the bibliographic file, which is known as Inpadoc, there is one record per patent document, and a set of special commands for linking records into the family group. A command is also available which will allow legal status data to be printed for a family, without needing to access Legstat separately. Orbit also has a current awareness file, Inpanew, which contains records for the last six weeks.

On Dialog the bibliographic and legal status data are available together in file 345. Complete data on each patent family member is available in one record.

On STN the file is mounted in a similar way to Dialog. STN also has Inpamonitor, containing records for the most recent six weeks.

World Patents Index. Produced by Derwent Publications. Currently covers 31 issuing authorities, but a number of these have been added only fairly recently. Each record relates to a patent family and currently there are approximately 5.5 million records. The file is split at 1980, but new equivalents are added to records on the back file. The most recent part of the database is known as World Patents Index Latest (often given as WPI/L) whilst the older part retains the title of World Patents Index (WPI). Earliest coverage is from 1963.

Gives brief bibliographic details and additional subject information (added by Derwent's subject-specialist indexers): classification, expanded title, abstract, index terms and additional words. (Extra coding is available to subscribers.) A large amount of documentation is available from Derwent. Available on Dialog, Orbit and Questel.

EDOC. Gives patent numbers, priority data and ECLA codes only. Covers 18 countries plus EP, PCT and OAPI applications. Earliest coverage from 1877 (Germany). Includes family information. Available on Questel.

Single-country files

Chinapats (Chinese Patent Abstracts in English). Chinese patent applications from 1985. Bibliographic information, including English language abstracts for applications filed by Chinese applicants. Equivalents information is included. Updated monthly.
Available on Orbit and Dialog.

EPAT. European Patent Office patents, published applications and Euro-PCT applications from 1978. Includes bibliographic, administrative and legal information and the text of the first claim of granted patents. Updated weekly, on day of publication.
Available on Questel.

FPAT. French patents from 1966. Includes bibliographic, administrative and legal information. Abstracts are available for most records from 1981, and English and French descriptors for records from 1987. Updated weekly, on day of publication.
Available on Questel.

ITALPAT. Covers Italian patent applications and Italian utility, design and ornamental model applications from 1983. The language of the database is Italian. Updated quarterly.
Available on ESA-IRS.

Japio. Japanese unexamined patent applications from 1976. Includes bibliographic data and English language abstracts.
Available on Orbit.

PATDPA. West German patent documents and utility models from 1968. Includes bibliographic, administrative and legal status data, and also covers EP and PCT applications designating Germany. Records for DE documents from 1981 include abstracts. The language of the database is German.
Available on STN.

PATOLIS. Japanese unexamined patents and utility models from 1971, examined patents from 1955 and examined utility models from 1961. The language of the database is Japanese (SRIS can help prepare searches carried out on this datebase by Epidos at Vienna, and to interpret the results).
Available from Japio.

USPA. US patents, from 1971. The file splits at 1981. Includes complete front page information, including the abstract, and the full text of all claims. Updated weekly, eight days after publication.
Available on Orbit.

Claims. The Claims family of files, produced by IFI/Plenum Data Corporation, cover most aspects of US patents.

Claims/US Patents Abstracts
Bibliographic data, including abstract and main claim, for US patents.

Earliest coverage 1950 (chemical only; mechanical and electrical from 1963). Design patents covered from 1980. Claims/US Patents Abstracts Weekly is also available, on Dialog, containing records for the current month only.

Claims/Class (Claims/Reference)
Dictionary index to the USPTO classification.

Claims/Citation
Patents cited by US patents, from 1947.

Claims/RRX
Information on US patents reassigned since 1980, reexamined since 1981, expired due to non-payment of fees since 1985, or extended since 1986.

Claims/Uniterm
As Claims/Patents with uniterm subject indexing for chemical patents.

Claims/Comprehensive
As Claims/Uniterm with additional subject indexing and search features.
Available by subscription only.

Claims/Compound Registry
Registry of Claims uniterms.

Available on Dialog, Orbit and STN. (Known on STN as the IFIClaims files)

USCLASS. US classification for all US patents from 1790.
Available on Orbit.

LEXPAT. Full text of US patents, from 1975.
Available on Mead Data Central: subscription needed.

PATDATA. US utility patents since 1971 and other patent publications from 1975.
Available on BRS.

Subject-specialised files

Apipat. Covers patents from 31 countries on petroleum refining and petrochemicals. Earliest coverage from 1964 (1964-1982 covered only 11 patent issuing authorities)

Available on Dialog and STN, and also on Orbit as part of WPIA and WPILA which also contains the whole of WPI and WPI/L. There are restrictions on access.

CA Search. The online version of Chemical Abstracts. Includes patents from 29 countries on all aspects of chemistry and chemical engineering. Earliest coverage from 1964.
Available on Data-Star, Dialog, ESA-IRS, Orbit, Questel and STN.

Pharmsearch. Covers EP and US patents in the pharmaceutical field since 1978 and French patents in this area since 1961. A companion file, MPHARM allows structural searching with Markush DARC.
Available on Questel.

Space patents. Covers patents in the area of space technology, from 1950. Contains c. 2500 records. Produced by the European Space Agency; updated occasionally.
Available on ESA-IRS.

Current Patents Evaluations. Covers new applications from Britain, the European Patent Office and Germany, and new American patents, relating to pharmaceuticals, agrochemicals and biotechnology.
Three files:

> *CPFN (Current Patents Fast New):* Most recent six weeks of data

> *CPFA (Current Patents Fast Alert):* 1989 until six weeks ago

> *CPEV (Current Patents Evaluated):* From 1990. Includes evaluations of applications considered to be 'patents of merit in the fields of pharmaceuticals and biotechnology'.

Available on Orbit, Data-Star.

Drug Patents International. Contains over 13,400 records of evaluated drug patent information from all major patenting countries, including comments on patent cover, priority and application data, CAS registry numbers, trade names and therapeutic use. Information on estimated expiry date is available only to subscribers to the hard copy equivalent.
Available on Orbit.

Patent litigation and other patent-related files

ECLATX. Full text of the European Patent Office classification scheme, ECLA.
Available on Questel.

Jurinpi. Published or unpublished French or EP jurisprudence concerning French patents. From 1823.
Available on Questel.

LEXIS. A group of full-text databases containing reports from US and British courts on patent and trade mark cases.
Produced by Butterworth but available on Mead Data Central: subscription needed.

Litalert. Includes notices of filing and subsequent action for patent infringement suits filed in US District Courts. From 1970.
Available on Orbit.

Merck Index. Full text of the Merck Index, 11th edition, 1989, covering pharmaceuticals, with updates twice a year. Includes references to relevant patent numbers.
Available on BRS, Dialog and Questel.

NTIS. Details of US government-sponsored research. The National Technical Information Service (NTIS) uses the database to announce patents and patent applications issued to US Government agencies which are available for licensing. From 1973.
Available on Orbit, STN, ESA-IRS, Data-Star, BRS and Dialog.

PAST (Patent Status). Contains over 211,000 records of post-issue actions affecting the status of US patents, including Certificates of correction, Expirations, Requests for re-examination and patent suits. From 1973.
Available on Orbit.

Tech-line. Analyses technological strength of 1100 companies patenting in the USA using American patent data. From 1983.
Available on Mead Data Central.

TRANSIN. Technology transfer proposals: technological innovations in all fields, available for licensing, or seeking an associate, distributor etc.
Available on Questel.

Compact disc databases

ACCESS. Bibliographic information including abstracts on EPO patent applications. From 1978 onwards on one disc. Updated four times annually. Available from Chadwyck-Healey or the EPO.

APS. Bibliographic information and abstracts of US patents, designs, plants etc. From 1969 onwards on five discs. Current disc updated monthly.
Available from Chadwyck-Healey.

BULLETIN. Bibliographic and status information on EPO patents. From 1978 onwards on one disc. Updated four times annually. Available from Chadwyck-Healey or the EPO.

CASSIS. Bibliographic information on US patents, designs, plants etc. In three discs: Bibliographic, containing bibliographic data for patents from 1969 and abstracts for the last three years (non-patents from 1977 only), plus the manual of classification; Classification, containing file lists from 1790; and Asist, containing many miscellaneous files. Updated three times annually.
Available from MicroInfo.

CLAIMS. Bibliographic data on US chemical patents from 1950 and mechanical/electrical patents from 1963. Updated bimonthly.
Available from SilverPlatter.

FIRST. Full text and illustration storage of PCT and EPO front pages with about 12,000 documents covered in each disc. Searchable by most bibliographic elements. From 1988, discs issued five times annually.
Available from Chadwyck-Healey or the EPO.

IPC:CLASS. 3rd, 4th and 5th editions of the International Patent Classification plus catchword indexes and concordances.
Available from WIPO.

OG-PLUS. Full text and illustration storage of the US *Official gazette* with data from Patent Status File and LitAlert databases. From July 1990, discs issued weekly.
Available from Research Publications.

This page intentionally left blank

11 SRIS AND THE PATENTS INFORMATION NETWORK

Background

SRIS (Science Reference and Information Service) is part of the London Services division of the British Library. It has inherited the collections of the Patent Office Library, which was founded in 1855. Today SRIS continues to develop its extensive collection of industrial property literature alongside its other information resources in the fields of science, technology and commerce.

SRIS is to move to the new British Library building at St Pancras in 1993.

Resources

Industrial property literature constitutes the largest single specialist resource within SRIS. About 31 million patent specifications are held from (currently) 35 countries or international systems. About one million documents are added annually, though these constitute only half a million inventions.

The specifications are arranged by country, then by publication stage, and then by number. Most of the specifications are held as paper, but an increasing number are held as microfiche or microfilm. This is mostly in the case of the smaller countries but sets duplicating the paper sets are also held for some large patenting authorities. CD-ROMs of patents are now beginning to be acquired as well.

The literature is held in two reading rooms and in various stores:

(1) SRIS Holborn Reading Room (25 Southampton Buildings)

The Holborn Reading Room holds the following literature:-

- British, European Patent Office and Patent Cooperation Treaty patents plus gazettes, classified abstracts, etc.

- Books and journals on industrial property.

- Non-official status registers.

- Inpadoc and Derwent microfiches indexes.

Older British patents and some older foreign material are kept in vaults on-site which can be visited.

Two unique card indexes (compiled from information supplied by the Patent Office) are maintained in the Reading Room. The Card index of applicants for British patents is kept in cumulative annual sequences and is used to trace recent applications. The Index of applicants for British designs has been maintained since 1961. The title and registration number of a design can be found from it and copies of the design can then be requested from the Designs Registry at the Patent Office.

Case law material is also held in the Reading Room. This includes journals giving accounts of court cases in the area of intellectual property as well as typescripts of recent Patent Office, High Court of Chancery and European Patent Office decisions. Card indexes to these series are available.

(2) Foreign Patents Reading Room (Chancery House)

This reading room (located across the street from 25 Southampton Buildings) holds the following material:-

- Foreign patents from 32 countries or systems.

- Gazettes from 88 countries or territories.

- Registers recording the date of receipt of patents.

- Inpadoc and Derwent microfiche indexes.

- Most of the Derwent abstracting journals.

Indexes to US patents are also held (on microfilm cassettes). These list all US patents from 1790 onwards (including reclassified patents) by US classification number.

CD-ROMs containing US patent data can be searched in the CD-ROM Search Room.

Most older foreign patents are held in an off-site store. A regular delivery service to the Reading Rooms is in operation.

Services

SRIS provides a range of services to users of its industrial property resources. These involve identifying, locating and supplying information or citations. A number of these services are provided free of charge, and a charge is made for others.

Free services

Both the Holborn Reading Room and the Foreign Patents Reading Room have patents enquiry desks. The specialist staff manning these desks are able to help visitors to the library carry out their searches in the patent literature (but they do not have the time to carry out searches for visitors).

Enquirers who telephone the desks (or who write or telex or fax an enquiry), will for example be given quick answers to the following sorts of questions:-

- Status of British patents.

- Information on British patent applications from the card index.

- Date of receipt of foreign patents.

- Equivalents of a particular patent.

Free leaflets and brochures, including those by the Patent Office, are produced and are available on request.

Priced services

Patents Online Search Service (POSS). SRIS has expert staff who can plan and carry out searches for clients using online and CD-ROM databases.

Seminars. SRIS runs a regular programme of seminars in the field of patent information.

Publications. SRIS publishes new and revised publications on patents and patent related topics.

Patent Express. this is the SRIS photocopy service. It has levels of service to meet a wide range of customer needs and accepts orders from occasional customers and from account holders.

FOR FURTHER INFORMATION TELEPHONE

Holborn Reading Room British Patents Desk	071 323 7919
Foreign Patents Reading Room Desk	071 323 7902
Seminars (SRIS Marketing)	071 323 7470
Publications	071 323 7472
Patents Online Search Service	071 323 7903
Patent Express (photocopies)	071 323 7926/8

Patents Information Network (PIN)

The Patents Information Network provides key patent documentation and related services from regional centres spread throughout the United Kingdom.

The PIN is made up of six Patent Information Libraries (which have substantial holdings of British and European PCT patent documentation) and seven Patent Information Centres (which have more limited holdings.)

The Centres have also developed a range of other services to meet local needs. These services may include photocopying facilities, rapid supply of documents, access to patents online databases, patents clinics offering free consultation with local patent agents, and meetings and workshops on patents and related topics.

**Patents Information
Libraries/Centres**

(★ = PIN Centre)

Aberdeen★
Central Library
Rosemount Viaduct
Aberdeen AB9 lGU
Tel 0224 634622

Belfast
Patents Section
Science Library
Central Library
Belfast Public Library
Belfast BT1 1EA
Tel 0232 24323

Birmingham
Patents Department
Central Library
Chamberlain Square
Birmingham B3 3HQ
Tel 021 2354537

Bristol★
Library of Commerce and Industry
Central Library
College Green
Bristol BS1 5TL
Tel 0272 299148

Coventry★
Lanchester Library
Coventry Polytechnic
Much Park Street
Coventry CV1 2HF
Tel 0203 838448

Glasgow
Patents Collection
Business Users' Service
Mitchell Library
North Street
Glasgow G3 7DN
Tel 041 221 7030

Leeds
Patents Information Unit
Leeds Public Libraries
32 York Road
Leeds LS9 8TD
Tel 0532 488747

Liverpool
Science and Technology Library
Central Libraries
William Brown Street
Liverpool L3 8EW
Tel 051 225 5442

Manchester★
Technical Library
Central Library
St Peter's Square
Manchester M2 5PD
Tel 061 234 1987

Newcastle upon Tyne
Patents Library
Central Library
Princess Square
Newcastle upon Tyne
NE99 1DX
Tel 091 261 0691

Plymouth★
Reference Department
Central Library
Drake Circus
Plymouth PL4 8AL
Tel 0752 385906

Portsmouth★
Central Library
Guildhall Square
Portsmouth PO1 2DX
Tel 0705 819311

Sheffield★
Science and Technology Library
Central Library
Surrey Street
Sheffield S1 1XZ
Tel 0742 734742

12 BIBLIOGRAPHY OF INDUSTRIAL PROPERTY

Scope

This compilation represents a selection of publications held by SRIS dealing with patents. The majority of the titles listed are those found to be currently consulted by users and staff in SRIS.

The inclusion of a title is no guarantee of its usefulness to every user of this list; nor does exclusion of a title reflect on its quality.

Arrangement of literature at SRIS

To help find any of the items on the shelves at SRIS, the location has been given at the foot of each entry. Items having a location mark prefix (B) will be found in the book sequence while those prefixed (P) are in the periodical sequence. Items located at 'British Patents' are in the patents area of the reading room in 25 Southampton Buildings and those at 'Foreign Patents' are in the nearby reading room in Chancery House. Details of the SRIS classification can be obtained from the enquiry desk.

Abbreviations used

CPC	Community Patent Convention
EP	European Patent
EPC	European Patent Office
PCT	Patent Cooperation Treaty
SRIS	Science Reference and Information Service (formerly SRL)
SRL	Science Reference Library
WIPO	World Intellectual Property Organisation

1. British and European legislation and its review

1.1. The British patent system. Report of the committee to examine the patent system and patent law. Chairman: Banks, M.A.L.
London: HMSO, 1970.
Cmnd 4407
SRIS Location: (B) BF 32

1.2. *Convention on the patent for the Common Market* (Community Patent Convention). Luxembourg, 15 December 1975.
London: HMSO.
Proposal for a single Community Patent valid in all EEC countries but not yet in force.
SRIS Location: See Luxembourg Conference on the Community Patent 1975 at (B) BG 32.

1.3. *Convention on the grant of European patents* (European Patent Convention). Munich, 5 October 1973. Miscellaneous No.24 (1974)
London: HMSO, 1974.
Cmnd 5656.
SRIS Location: (B) BG 301

1.4. *Copyright, Designs and Patent Act 1988.*
London: HMSO, 1988.
Alters mainly copyright and design laws.
SRIS Location: (B) BF 01.

1.5. *Intellectual property rights and innovation.* London: HMSO, 1983.
Cmnd. 9117.
Discussion document (Green Paper) prepared by the Cabinet Office, recommending changes relating to the Patent Office, the role of Government and the abuse of rights.
SRIS Location: (B) BF 02.

1.6. *Intellectual property and innovation.*
London: HMSO, 1986.
Cmnd 9712.
The Government's White Paper containing proposals for reform of patents, designs and copyright following 1.5. and subsequent Green Papers on copyright, design and performers' protection and the recording and rental of audio and video copyright material.
SRIS Location: (B) BF 02.

1.7. *Patent Cooperation Treaty (with Regulations).* Washington, 1970.
London: HMSO, 1978.
Treaty Series No. 78 (1978). Cmnd 7340.
SRIS Location: British Patents.

1.8. *Patents Act 1977.*
London: HMSO, 1977.
The UK Act of Parliament governing the present patent systems in the UK which came into effect on 1st June 1978.
SRIS Location: (B) BF 31.

1.9. *The texts established by the Luxembourg Conference on the Community patent 1985.*
Luxembourg: Council of the European Communities,1986.
Resolution of some of the outstanding difficulties for the CPC (see 1.2.).
SRIS Location: (B) BG 32.

2. British and European Patent Office patent law

2.1. Aldous, W. et al. *Terrell on the law of patents.* 13th ed.
London: Sweet and Maxwell, 1982.
After a short history and discussion of the nature of patentable inventions, the 1977 and the surviving parts of the 1949 UK Patents Acts are appraised. The effect of the EPC is considered and the appendices give the texts of the EPC, PCT, rules of the Supreme Court, articles from the Treaty of Rome, as well as the 1977 and 1949 UK Patent Acts.
SRIS Location: (B) BF 30.

2.2. Baxter, J.W. *World patent law and practice.*
London: Sweet and Maxwell; New York: Matthew Bender, 1973-.
A multinational reference work for patent practitioners arranged in subject order so that the position in any topic can be easily obtained for a large number of countries. Additional chapters contain details of conventions and new legislation. Loose-leaf updated about three times each year.
SRIS Location: (B) BH 30.

2.3. Blanco White, T.A. *Patents for inventions.* 5th ed.
London: Stevens & Sons, 1983.
Commentary on the UK 1949 Patents Act and the effect of the 1977 Act upon it.
SRIS Location: (B) BF 30.

2.4. Blanco White, T.A. et al. *The encyclopedia of United Kingdom and European patent law.*
London: Sweet and Maxwell, 1977-.
Contains over 300 pages of commentary and includes the UK 1949 and 1977 Patent Acts with Rules; EPC, CPC and PCT with Rules, the Paris Convention for the Protection of Industrial Property first signed in 1883 and the Strasbourg Conventions of 1963 and 1971 on the unification of patent law and classification. In loose-leaf format and updated periodically.
SRIS Location: (B) BG 30.

2.5. *Butterworths intellectual property handbook.* Consulting editor, J. Phillips.
London: Butterworths, 1990.
Reprints numerous laws and statutory instruments.
SRIS Location: (B) BF 01.

2.6. Chartered Institute of Patent Agents. *CIPA guide to the Patents Act 1977.* 3rd ed.
London: Sweet and Maxwell, 1990-.
Reproduces the UK Act section by section, with commentary. Rules made under the Act and other statutes are added as an appendix. Supplements provide comments on cases and more recently published rules.
SRIS Location: (B) BF 30.

2.7. Chartered Institute of Patent Agents. *European patents handbook*. 2nd ed.
London: Oyez Longman, 1988-.
Contains over 350 pages of commentary and reproduces EPC, CPC, PCT texts and guidelines for examination under the EPC and PCT. Loose-leaf, updated periodically.
SRIS Location: (B) BG 30.

2.8. Cawthra, B.I. *Patent licensing in Europe*. 2nd ed.
London: Butterworths, 1986.
Traces recent developments of the patent licensing system in Europe citing significant decisions of the European Court and provides commentary on clauses in licensing agreements related to EEC countries giving some comparison with American law.
SRIS Location: (B) BH 362.

2.9. Phillips, J. and Firth, A. *Introduction to intellectual property law*. 2nd ed.
London: Butterworths, 1990.
Readable introduction to basic intellectual property principles.
SRIS Location: (B) BF 00.

2.10. Robertson, R. *Legal protection of computer software*.
London: Longman, 1990.
Covers the protection available for software by means of patents as well as by laws of copyright, trade secrets and trade marks. Also covers remedies.
SRIS Location: (B) BF 191.

2.11. *The taxation of patent royalties, dividends, interest in Europe*.
Amsterdam: International Bureau of Fiscal Documentation. 1963-. A Guide to 18 countries, loose-leaf updated about once a year.
SRIS Location: (B) BH 12.

2.12. Walton, A.M. and Laddie, H.I.L. et al. *Patent law of Europe and the United Kingdom*.
London: Butterworths, 1981-2.
Contains commentary on the UK 1949 and 1977 Acts, EPC, CPC and PCT and reproduces the UK statutes, the EPC, CPC and the PCT.
Other sections contain information on forms and precedents, practice etc.
SRIS Location: (B) BG 30.

3. United States patent law

3.1. Chisum, D.D. Patents: *a treatise on the laws of patentability, validity and infringement*.
New York: Matthew Bender, 1978.
Highly detailed analysis of American laws with extensive reference to court cases. Includes an index of the cases referred to. Loose-leaf updated periodically.
SRIS Location: (B) BL 36.

3.2. Lipscomb, E.B. *Lipscomb's Walker on patents*, 3rd ed.
Rochester, N.Y.: Lawyers Co-operative Publishing, 1984-.
Discusses designs, plants etc. as well as patents with extensive reference to
court cases. Includes an index of the cases referred to. Bound, with
loose-leaf updates periodically.
SRIS Location: (B) BL 30.

3.3. *United States Code Annotated, Title 35, Patents.*
St Paul, Minn.: West Publishing, 1984.
Includes text of Patent Office regulations, etc. and PCT matters.
Extensive references to court cases.
SRIS Location: (B) BL 31.

4. German patent law

4.1. *Gebrauchsmustergesetz 1987.* Ed. by H-F. Klunker et al.
Cologne: Heymanns, 1986.
Consists of German text, with parallel English translation, of the German
Utility Model Act 1987.
SRIS Location: (B) BJ 29.

4.2. *Industrial property laws of the Federal Republic of Germany: patent, utility
model, trademark, design.* Ed. by V. Vossius and U.C. Hallmann. 2nd ed.
Munich: Wila, 1985.
Consists of texts of the laws.
SRIS location: (B) BJ 25.

4.3. Krasser, R. *Lehrbuch des Patentrechts.* 4th ed.
Munich: Beck'sche, 1986.
Discusses, in German, German and European Patent Office laws.
SRIS Location: (B) BJ 29.

5. Japanese patent law

5.1. Patent Office. *Examination manual for patent and utility model in Japan.*
Tokyo: AIPPI, 1986-.
In English. A guide to examining procedure. Loose-leaf updated
periodically.
SRIS Location: (B) BK 04.

5.2. Patent Office. *Guide to industrial property in Japan.*
Tokyo: Patent Office, 1988.
In English. A useful guide to application and examination procedures in
Japan.
SRIS Location: (B) BK 00.

6. Obtaining patent protection

6.1. Crespi, R.S. *Patenting in the biological sciences.*
Chichester: Wiley, 1982.
An introduction to patenting procedures for research workers in
biotechnology and the pharmaceutical and agrochemical industries.
SRIS Location: (B) BH 45.

6.2. Eisenschitz, T.S. and Phillips, J. *The inventor's information guide.* 2nd ed. London: Queen Mary College, University of London, 1985. Aimed at the individual inventor who lacks the resources of professionally employed inventors, it covers information sources, acquiring a patent, developing an invention, professional services, finance, marketing, designs, addresses and publications.
SRIS Location: (B) BF 21.

6.3. Greene, Ann Marie. *Patents throughout the world.*
New York: Trade Activities Inc., 1981-.
Covers the procedures of over 150 countries and includes sections on international conventions; entries are fairly brief but are updated about three times a year. Loose-leaf format.
SRIS Location: British Patents and Foreign Patents.

6.4. Grubb, P.W. *Patents in chemistry and biotechnology.*
Oxford: Oxford University Press, 1986.
A guide for the chemist inventor providing sections on patenting, the history of patents and the effect of law in the EEC, USA, Comecon and developing countries. It also contains a glossary of patent terms.
SRIS Location: (B) BH 30.

6.5. *Guidelines for examination in the European Patent Office.*
Munich: EPO, 1992.
Contains sections on formalities, search, substantive examination, opposition, etc. Loose-leaf updated periodically.
SRIS Location: British Patents.

6.6. *How to get a European patent.* 8th ed.
Munich: EPO, 1989.
A guide for applicants giving general information on the procedure.
SRIS location: British Patents.

6.7. *How to prepare a UK patent application.*
Newport: Patent Office, undated.
A layman's guide to the procedure for applying for a UK patent under the 1977 Act.
Free of charge from the Patent Office.

6.8. *Manual of patent practice in the UK Patent Office.* 2nd ed.
London: Patent Office, 1991.
Reproduces, with extensive commentary for the benefit of staff interpreting the law, the text of the 1977 Patents Act.
SRIS Location: British Patents.

6.9. *Patent protection.*
Newport: Patent Office, undated.
An introduction to obtaining patent protection under the EPC, PCT or UK national law.
Free of charge from the Patent Office

6.10. Katzarov, K. *Manual on industrial property.* 9th ed.
Geneva: Katzarov, 1981-.
Provides, in two volumes, the main features of industrial property
legislation and procedure in more than 150 countries; contains a section
on international conventions. Loose-leaf format, updated periodically.
SRIS Location: (B) BH 00.

6.11. *Manual for the handling of applications for patents, designs and trade marks
throughout the world.*
Amsterdam: Manual Industrial Property BV, 1936-.
Commonly referred to as 'The Dutch Manual'. Loose-leaf, updated by
supplements, in more detail than 6.3. but updated less frequently, about
once a year.
SRIS Location: British Patents; Foreign Patents

6.12. *National law relating to the EPC.* 3rd ed. Munich: EPO, 1983.
A comparison of the laws of member countries of the EPC summarised in
tabular form. Written as a guide to applicants for European patents. Free
of charge from the EPO.
SRIS Location: British Patents.

6.13. *PCT applicant's guide.*
Geneva: WIPO, 1991-.
Information on how to file applications, processing procedures, etc. in
three volumes. Loose-leaf, updated periodically.
SRIS Location: British Patents.

6.14. Shaw, L. *The practical guide for people with a new idea.* 2nd ed.
Birmingham: L. Shaw, 1990.
A brief guide to the protection of intellectual property for the inventor
wishing to exploit his idea.
SRIS Location: (B) BF 48.

7. Patents as information

7.1. *Commission of the European Communities. Patent information and
documentation in Western Europe.* 3rd ed. Edited by B M Rimmer.
Munich: K.G. Saur, 1988.
An inventory of material and services available to the public in 17
countries in Western Europe with additional sections on Japan, USA,
USSR, international organisations, associations, journals and standards for
patent literature.
SRIS Location: (B) BH 39.

7.2. Eisenschitz, T. *Patents, trade marks and designs in information work.*
London: Croom Helm, 1987.
Describes documentation and searching methods in industrial property.
SRIS Location: (B) BH 00.

7.3. *Industrial property publications in SRL - UK, European Patent Office and Patent Cooperation Treaty.*
London: British Library, Science Reference Library, 1985.
A guide to SRIS holdings of patents, trade marks and designs.
Free of charge from SRIS.

7.4. Kase, Francis J. *Foreign patents.*
Dobbs Ferry, New York: Oceana Publications; Leiden: Sijthoff, 1972.
A guide to the official literature published before 1972 by about 60 countries with reproductions of their publications.
SRIS Location: (B) BH 30.

7.5. *Handbook on industrial property information and documentation.*
Geneva: WIPO, 1990-.
In English and French. Contains sections on PCT minimum documentation; standards for patent documentation; International Patent Classification; PCT; kinds of patent publications (with samples of front pages); storage and copying (including a catalogue of microfilms); and access to patent documentation (listing many commercial and other services available). Loose-leaf format, two volumes periodically updated.
SRIS Location: British Patents and Foreign Patents.

7.6. Rimmer, B.M. *International guide to official industrial property publications.* 2nd ed.
London: British Library, 1988-.
Describes the specifications, gazettes and other publications of 40 patenting authorities with additional information on designs and trade marks. In loose-leaf format.
SRIS Location: (B) BH 39.

7.7. *World directory of sources of patent information.*
Geneva: WIPO, 1985.
Lists the information centres which collect patent publications giving details of the services, accessibility to the public and holdings available. An index provides a list of patent documents and the libraries' holdings of each.
SRIS Location: British Patents and Foreign Patents

7.8. *World patent information.*
McLean, Va.: Pergamon, 1979-.
Quarterly.
Produced jointly by the Commission of the European Communities and WIPO. Contains topical articles, notices on regulations and events, and reviews of new books and articles in periodicals.
SRIS Location: (P) BH 39 - E(l).

8. Invention; patenting history

8.1. *American enterprise: nineteenth-century patent models.*
New York: Cooper-Hewitt Museum, 1984.
Illustrates numerous models, arranged by topic, from the museum's collection.
SRIS Location: (B) BF 482.

8.2. Baker, R. *New and improved.*
London: British Library Publications, 1976.
Illustrated descriptions of 363 significant patented inventions dating from
1691 to 1971, e.g. typewriter, Portland cement, jet engine. Includes
patent numbers.
SRIS Location: (B) BF 482.

8.3. Davenport, Neil. *The United Kingdom patent system.*
Havant: Kenneth Mason, 1979.
A compact summary which contains much detail illustrating the
development of the UK patent system over 400 years up to the 1977 Act.
The bibliography includes a list of statutes and rules, law reports and
official publications with reports of Committees of Enquiry.
SRIS Location: (B) BF 46.

8.4. Desmond, K. *The Harwin chronology of inventions, innovations, discoveries.*
London: Constable, 1987.
Lists several thousand inventions in a chronological arrangement.
SRIS Location: (B) BF 48.

8.5. Dutton, H.I. *The patent system and inventive activity during the industrial
revolution, 1750 - 1852.* Manchester: Manchester University Press, 1984.
An assessment of the part played by patents in the development of
industry in Britain. Part I discusses the patent institution - the system,
law, courts, and patent agents. Part II reviews the economics of patents
and invention. Contains an extensive bibliography.
SRIS Location: (B) BF 46.

8.6. Dummer, G.W. *Electronic inventions and discoveries.* 3rd ed.
Oxford: Pergamon Press, 1983.
Lists 450 significant inventions from 1642 with source of information, and
sometimes a patent number.
SRIS Location: (B) RK 10.

8.7. Giscard d'Estaing, V.- A. *The world almanac book of inventions.*
New York: World Almanac Publications, 1985.
Lists 2000 inventions by subject.
SRIS Location: Centre Desk (B) BF 482.

8.8. Jewkes, J. et al. *The sources of invention.*
London: Macmillan, 1969.
A frequently cited book about inventors and inventions including more
than 50 case histories of important inventions, e.g. automatic transmission,
DDT, transistors.
SRIS Location: (B) BF 48.

8.9. Macleod, C. *Inventing the Industrial Revolution: the English patent system
1660-1800.*
Cambridge: CUP, 1988.
A detailed study of British patenting patetrns before 1800. Includes a
lengthy bibliography.
SRIS Location: (B) BF 46.

8.10. Newton, D.C. *New manufactures within this realm.*
London: British Library, Science Reference Library, 1985.
A chronological list of the principal changes affecting the law on patents in
the UK since the Statute of Monopolies 1624.
SRIS Location: (B) BF 46.

8.11. *The Paris Convention for the Protection of Industrial Property 1883 to
1983.*
Geneva: WIPO, 1983.
An account of the history of the Paris Convention and the evolution of
the governing secretariat and WIPO.
SRIS Location: (B) BG 205.

9. Case law

9.1. *Chartered Institute of Patent Agents. European patents sourcefinder.*
London: Longman, 1988-.
Gives details, with indexes, of European Patent Office decisions. Updated
at intervals.
SRIS location: British Patents.

9.2. *Fleet Street reports of industrial property cases from the Commonwealth and
Europe (previously Fleet Street patent law reports).*
London: European Law Centre. 1963-.
Commercially published edited transcripts of selected proceedings in
various courts. Some Patent Office decisions are included and also, since
1973, decisions of the European Court and European Commission.
SRIS Location: British Patents.

9.3. Fletcher-Moulton, H. *Digest of the patent, design trade mark and other
cases.*
London: Patent Office, 1959.
Digests of cases between 1883 and 1955 published in Reports of patent
cases .... (see 9.7.) with indexes by topic and party. Effectively continues
9.5, and continued in 9.4.
SRIS Location: British Patents.

9.4. Fysh, M. *The industrial property citator.*
London: European Law Centre, 1982-.
Indexes decisions in published Commonwealth law reports from 1955 by
topic and party. Updated periodically by supplements. Effectively
continues 9.3.
SRIS Location: British Patents.

9.5. Hayward, P.A. *Hayward's Patent cases 1600-1883.*
Abingdon: Professional Books, 1987.
Reprints and indexes in 11 volumes published patent decisions from
various sources. Effectively continued by 9.3.
SRIS Location: British Patents.

9.6. *Intellectual property decisions.*
Sudbury: Centre for Legal and Business Information, 1978-.
10 issues p.a.
Provides alerting abstracts of most of the significant decisions in British
and European courts in the field of intellectual property.
SRIS Location: British Patents.

9.7. *Reports of patent, design and trade mark cases.*
London: The Patent Office, 1884-.
An officially printed series compiled by a legal editor giving the
judgements delivered by various courts in selected cases.
SRIS Location: British Patents.

10. Current reviews

10.1. *European intellectual property review.*
Oxford: ESC Publishing, 1978-.
Monthly.
Articles of current interest and a section on recent legal cases, legislation
and news items, country by country.
SRIS location: (P) BH 00 - E(6).

10.2. *Industrial property.*
Geneva: WIPO, 1962-.
Monthly.
A primary source of official international information with occasional
newsletters from patenting authorities and a pull-out section on new
legislation. The January issue of each year lists the members of numerous
intellectual property treaties etc.
SRIS location: (P) BH 00 - E(5).

See also 7.8.

11. Statistics

11.1. *Industrial property statistics.*
Geneva: WIPO, 1976-.
Annual.
Provides statistics on all sections of industrial property activity in WIPO
member countries.
SRIS location: (P) BH 10 - E(l).

11.2. *100 years protection of industrial property: statistics.*
Geneva: WIPO, 1983.
Synoptic tables on patents, trade marks, designs, utility models and plant
varieties 1883-1982.
SRIS location: (P) BH 10 - E(2).

12. Dictionaries and glossaries

12.1. Berson, A.S. et al. *English-Russian patent dictionary.*
Moscow: Soviet Encyclopedia Publishing House, 1973.
Includes lists of expanded and translated abbreviations and translations of titles of official gazettes.
SRIS location: (B) AA 145.

12.2. *Industrial property glossary.*
Geneva: WIPO.
A series of publications giving translations of about 300 main terms some of which are subdivided into related terms:

    English-French-Portuguese   (1980)
    English-French-Russian   (1980)
    English-French-Chinese   (1981)
    English-French-German   (1982)
    SRIS location: (B) AA 161.

12.3. Kase, Francis J. *Dictionary of industrial property terms: English, Spanish, French, German.*
Alphen van den Rijn: Sijthoff and Noordhoff, 1980.
SRIS location:(B) AA 161.

12.4. Klaften, B., Wittman, A. and Klos, J. *Worterbuch der Patentfachsprache = Patent terminological dictionary, English-German, German-English.* 5th ed.
Munich: Wila, 1986.
Contains a supplement of terms relating to patent drawings.
SRIS location: (B) AA ll9(GER).

12.5. Russell, Robert W. (compiler). *Patents and trademarks in Japan.* 3rd ed.
Tokyo: Russell, 1974.
More than 700 Japanese terms transliterated and arranged in alphabetical orders of English meaning with a full explanation of their significance in Japanese law.
SRIS location: (B) BK 00.

12.6. Szendy, Gyorgy. *Worterbuch des Patentwesens in funf Sprachen.* 2nd ed.
Dusseldorf, VDI Verlag, 1985.
German, English, French, Spanish, Russian.
SRIS location: (B) AA 161.

12.7. Ueki, Eikichi. *Six-Languages dictionary of industrial properties.*
Tokyo: Patent Data Centre, 1979.
Japanese, English, French, German, Russian, Spanish.
SRIS location: (B) AA 161.

12.8. Uexkull, J-Detlev Von, and Reich, H.J. Worterbuch der Patentpraxis.
Koln: Heymanns, 1983.
German-English, English-German.
SRIS location: (B) AA 119.

12.9. Work, H., et al. *English-Russian-Estonian patent dictionary.*
Tallinn: Valgus, 1976.
Translations are given for words, associated terms and phrases encountered
in connection with patents.
SRIS location: (B) AA 161.

This page intentionally left blank

13 GLOSSARY OF PATENT TERMS

This glossary of definitions gives 'simplified' rather than 'official' definitions. Some of these words are defined in somewhat different forms across the world, and sources listed in the bibliography should be consulted for variations. All the words in bold are defined in this list.

Anticipation. When the **prior art** indicates that a patent application lacks **novelty**.

Applicant. The person or corporate body that applies for the **patent** and intends to 'work' the invention, i.e. to manufacture or licence the technology.

Assignee. In the USA, the person or corporate body who acquires the rights to manufacture or license an invention from the inventor, often by contingency contracts.

Claims. The definition at application of the monopoly that the **applicant** is trying to obtain for the invention, or the actual monopoly that is given at **grant**.

Convention. 'Filing by the Convention' means obeying the rules established by the 1883 Paris Convention. This usually has the connotation of filing foreign applications within 12 months of the **priority date** application.

Disclosure. The first publication of details of an invention. This may be deliberately revealed outside the patent system to make the invention unpatentable.

Disposal. A term used in some countries such as the USA to mean that an application has been resolved by being withdrawn, rejected or granted. It can also have the connotation of being rejected only.

Equivalent. These are **specifications** published by different patent offices for the same invention. Together they form the patent family.

Examination. See **Preliminary examination** and **Substantive examination**.

Expiry. The date when a **patent** has run its full **term** in a country and is no longer protected in there. Can also be used to mean **lapsing**.

Grant. A temporary right given by a patent office to an **applicant** to prevent anyone else from using the technology defined in the **claims** of a **patent**.

Infringement. An alleged or actual manufacture or import of an invention currently protected by a **patent**.

Lapsing. The date when a **patent** is no longer protected in a country or system due to failure to pay **renewal fees**. Often, however, the patent can be **reinstated** within a limited period.

Maintenance fees. *See* **Renewal fees**

Novelty. The concept that the **claims** defining an invention in a patent application must be totally new. Most patent offices define this as not being revealed or publicly available anywhere in the world before the **priority date** but in the USA novelty is normally determined by the date of invention.

Obviousness. The concept that the **claims** defining an invention in a patent application must not be a predictable improvement on what has been done or published before the **priority date**.

Open to public inspection (OPI). The date when a patent application was first made available to the public to see. This is normally not less than 18 months from the **priority date** but patent offices vary in their treatment.

Opposition. A request to the patent office by an opposing party that an application should be refused, or that a granted patent should be annulled.

Patent. Document defining rights conferred by the **grant**, but often used to mean any published **specification**.

Patent family. All the **equivalents** of a **specification**.

Patentability. The ability of an invention to satisfy the legal requirements for obtaining a **patent**, including **novelty**. Some types of inventions, e.g. computer software and plants, may be unpatentable in many countries.

Patentee. In the USA, the inventor in a **specification**, who has the theoretical rights to the invention, although they are often signed over to an **assignee**.

Pending. When a patent office has not yet decided whether to **reject** or to **grant** a patent application, and it has not been **withdrawn**.

Preliminary examination. The initial study of an application by a patent office, which in Britain involves checking that the **specification** is properly set out, and preparing a **search report**.

Prior art. Previously used or published technology, that may be referred to in an application.

Priority date. The initial date of filing of a patent application, normally in the applicant's domestic patent office. This date is used to help determine the **novelty** of an invention.

Reinstatement. Restoring a **patent** to protection after it has apparently **lapsed** by error or been revoked.

Rejection. When a patent office decides to refuse a patent application on one or more grounds.

Renewal fees. Payments that must be made by the **applicant** to keep the patent in force and prevent it from **lapsing**. Called maintenance fees in the United States.

Revocation. Stopping the protection given to a **patent** because of e.g. lack of **novelty**.

Search report. The list of citations of published **prior art** documents prepared by the patent office examiner in checking the **novelty** of a patent application.

Specification. The description, drawings and claims of an invention prepared to support a patent application. The term does not imply that the invention is necessarily new or was ever protected.

Status. The legal standing of a patent or patent application i.e. pending, lapsed, still protected.

Substantive examination. The examination by the patent office examiner of a patent application to determine whether a **patent** should be **granted**.

Term of patent. The maximum number of years that the monopoly rights conferred by the **grant** of a **patent** may last.

Utility model. A kind of patent available in some countries which involves a simpler inventive step than that in a patent. Also known as a petty patent.

Withdrawn. The permanent abandonment of a patent application either before or after publication. Often used to mean rejection by the patent office as well as withdrawal by the **applicant**.

This page intentionally left blank

14 INID AND COUNTRY CODES

INID codes are internationally agreed numbers for the identification of bibliographic data, hence the acronym.

Many patenting authorities, including the UK Patent Office, tag each item on the front page of their patent specifications with a number in brackets or a circle, so that the item can be identified irrespective of the language used. The following is a simplified list. Fuller details are in other sources such as 7.6 of the bibliography.

In a similar way country codes are frequently cited on patent documents and when indicating countries on databases.

INID codes

(10) Identification of the publication

(11) Number of the publication

(12) Kind of the publication

(19) Country code, or other identification, of the country of publication

(20) Local filing details

(21) Number given to the application

(22) Date of making application

(23) Other date(s) of filing

(24) Date from which industrial property rights may have effect

(30) Priority details

(31) Number assigned to priority application

(32) Date of filing of priority application

(33) Country in which priority application was filed.

> Note: International convention allows an application to be made in any country as if it had been made at the same time as the original (priority) application.

(40) Date of publication

41) Date of making available to the public by viewing, or copying on request, an unexamined specification which has not yet been granted

(42) Date of making available to the public by viewing, or copying on request, an examined specification which has not yet been granted

(43) Date of publication by printing of an unexamined specification which has not yet been granted

(44) Date of publication by printing of an examined specification which has not yet been granted

(45) Date of publication by printing of a granted patent

(46) Date of publication by printing of the claim(s) only

(47) Date of making a granted patent available to the public by viewing, or copying on request

(50) Technical information

(51) International Patent Classification

(52) Domestic or national Classification

(53) Universal Decimal Classification

(54) Title of the invention

(55) Keywords

(56) List of prior art documents

(57) Abstract or claim

(58) Field of search

(60) Reference to other legally related domestic document(s)

(61) Related by addition

(62) Related by division

(63) Related by reissue

(70) Identification of parties

(71) Name of applicant

(72) Name of inventor

(73) Name of grantee

(74) Name of attorney or agent

(75) Name of inventor who is also applicant

(76) Name of inventor who is also applicant and grantee

(80) Identification of data related to International Conventions

(81) Designated State(s) according to the Patent Cooperation Treaty (PCT)

(82) Elected State(s) according to the PCT

(84) Designated Contracting States under the European Patent Convention

(85)   Date of supply of the international patent application to the European Patent Office

(86)   Filing data of the international application

(87)   Publication data of the international application

(88)   Date of deferred publication of the search report

Country codes

Afghanistan	AF	Chile	CL
Albania	AL	China	CN
Algeria	DZ	Colombia	CO
Andorra	AD	Comoros	KM
Angola	AO	Congo	CG
Anguilla	AI	Costa Rica	CR
Antigua and	AG	Cuba	CU
Barbuda		Cyprus	CY
Argentina	AR	Czechoslovakia	CS
Aruba	AW		
Australia	AU	Denmark	DK
Austria	AT	Djibouti	DJ
		Dominica	DM
Bahamas	BS	Dominican	DO
Bahrain	BH	Republic	
Bangladesh	BD		
Barbados	BB	Ecuador	EC
Belgium	BE	Egypt	EG
Belize	BZ	El Salvador	SV
Benin	BJ	Equatorial Guinea	GQ
Bermuda	BM	Ethiopia	ET
Bhutan	BT		
Bolivia	BO	Fiji	FJ
Botswana	BW	Finland	FI
Brazil	BR	France	FR
British Virgin	VG		
Islands		Gabon	GA
Brunei	BN	Gambia	GM
Bulgaria	BG	German	DD
Burkina Faso	BF	Democratic Republic	
Burma	BU	German Federal	DE
Burundi	BI	Republic	
Byelorussian SSR	BY	Ghana	GH
		Gibraltar	GI
Cameroon	CM	Greece	GR
Canada	CA	Grenada	GD
Cape Verde	CV	Guatemala	GT
Cayman Islands	KY	Guernsey	GG
Central African	CF	Guinea	GN
Republic		Guinea-Bissau	GW
Chad	TD	Guyana	GY

Haiti	HT	Netherlands	NL
Honduras	HN	New Zealand	NZ
Hong Kong	HK	Nicaragua	NI
Hungary	HU	Niger	NE
		Nigeria	NG
Iceland	IS	Norway	NO
India	IN		
Indonesia	ID	Oman	OM
Iran	IR		
Iraq	IQ	Pakistan	PK
Ireland	IE	Panama	PA
Israel	IL	Papua New Guinea	PG
Italy	IT	Paraguay	PY
Ivory Coast	CI	Peru	PE
		Philippines	PH
Jamaica	JM	Poland	PL
Japan	JP	Portugal	PT
Jordan	JO		
		Qatar	QA
Kampuchea	KH		
Kenya	KE	Romania	RO
Kiribati	KI	Rwanda	RW
Korea, Democratic	KP		
People's Republic of		St Lucia	LC
Korea, Republic of	KR	St Helena	SH
Kuwait	KW	St Kitts-Nevis	KN
		St Vincent and Grenadines	VC
Laos	LA	Samoa	WS
Lebanon	LB	San Marino	MS
Lesotho	LS	Sao Tome and Principe	ST
Liberia	LR	Saudi Arabia	SA
Libya	LY	Senegal	SN
Liechenstein	LI	Seychelles	SC
Luxembourg	LU	Sierra Leone	SL
		Singapore	SG
Macao	MO	Solomon Islands	SB
Madagascar	MG	Somalia	SO
Malawi	MW	South Africa	ZA
Malaysia	MY	Soviet Union	SU
Maldives	MV	Spain	ES
Mali	ML	Sri Lanka	LK
Malta	MT	Sudan	SD
Mauritania	MR	Suriname	SR
Mauritius	MU	Swaziland	SZ
Mexico	MX	Sweden	SE
Monaco	MC	Switzerland	CH
Mongolia	MN	Syria	SY
Montserrat	MS		
Morocco	MA	Taiwan	TW
Mozambique	MZ	Tanzania	TZ
		Thailand	TH
Namibia	NA	Togo	TG
Nauru	NR	Tonga	TO
Nepal	NP	Trinidad and Tobago	TT

Tunisia	TN
Turkey	TR
Turks and Caicos Islands	TC
Tuvalu	TV
Uganda	UG
Ukrainian SSR	UA
United Arab Emirates	AE
United Kingdom	GB
United States of America	US
Uruguay	UY
Vanuatu	VU
Vatican City State	VA
Venezuela	VE
Vietnam	VN
Yemen	YM
Yugoslavia	YU
Zaire	ZR
Zambia	ZM
Zimbabwe	ZW

International organizations:

ARIPO	AP
Benelux	BX
European Patent Office	EP
OAPI	OA
Patent Cooperation Treaty	WO

This page intentionally left blank

INDEX

This page intentionally left blank

NOTES

Printed by Hobbs the Printers of Southampton